**Biological Monitoring
for Environmental Effects**

Biological Monitoring for Environmental Effects

Edited by
Douglas L. Worf
North Carolina
State University

LexingtonBooks
D.C. Heath and Company
Lexington, Massachusetts
Toronto

081447

Library of Congress Cataloging in Publication Data

Main entry under title:

Biological monitoring for environmental effects.

Includes index.
1. Indicators (Biology) 2. Environmental monitoring. I. Worf, Douglas L. [DNLM: 1. Environmental pollution—Analysis—Congresses. 2. Monitoring, Physiologic—Congresses. WA670 B615 1978]
QH541.15.I5B57   628.5'028   79-2977
ISBN 0-669-03306-5

Copyright © 1980 by D.C. Heath and Company

All rights reserved. No part of this publication may be reproduced or transmitted in any form or by any means, electronic or mechanical, including photocopy, recording, or any information storage or retrieval system, without permission in writing from the publisher.

Published simultaneously in Canada

Printed in the United States of America

International Standard Book Number: 0-669-03306-5

Library of Congress Catalog Card Number: 79-2977

Contents

	Foreword	ix
	Preface and Acknowledgments	xi
Part I	Introduction and Overviews	1
Chapter 1	Introduction *Joab Thomas*	3
Chapter 2	A Review of Environmental Data and Monitoring *John D. Buffington*	5
Chapter 3	Scenarios on Alternative Futures for Biological Monitoring, 1978-1985 *John Cairns, Jr.*	11
Part II	Monitoring Bioeffects of Water Pollution	23
Chapter 4	Federal and State Biomonitoring Programs *Cornelius I. Weber*	25
Chapter 5	Biological Monitoring to Provide an Early Warning of Environmental Contaminants *Kenneth L. Dickson, David Gruber, Christine King,* and *Kenneth Lubenski*	53
Chapter 6	Industrial Applications of Biological Monitoring in the Laboratory and Field *Alan W. Maki*	75
Chapter 7	Biomonitoring of Coastal Waters—An Overview *John P. Couch, Frank G. Lowman,* and *Ford A. Cross*	93
Chapter 8	Use of Benthic Macroinvertebrates as Indicators of Environmental Quality *D.R. Lenat, L.A. Smock,* and *D.L. Penrose*	97
Chapter 9	Research Suggestions—Benthic Invertebrates as Biological Indicators *John C. Morse*	113
Part III	Terrestrial Plant and Soil Biomonitoring	115
Chapter 10	Vegetation—Biological Indicators or Monitors of Air Pollutants *H.C. Jones* and *W.W. Heck*	117

Chapter 11	Biological Monitoring Techniques for Assessing Exposure *G.B. Wiersma, R.C. Rogers, J.C. McFarlane,* and *D.V. Bradley, Jr.*	123
Part IV	*Ecological Monitoring*	135
Chapter 12	Monitoring Cause and Effects—Ecosystem Changes *Allan Hirsch*	137
Chapter 13	National Environmental Research Parks: A Framework for Environmental Health Monitoring *William S. Osburn*	143
Chapter 14	The National Biological Monitoring Inventory *Robert L. Burgess*	153
Part V	*Bioassay of Environmental Pollution*	167
Chapter 15	Short-Term Bioassays of Environmental Samples *Michael D. Waters* and *Linda W. Little*	169
Chapter 16	Plants as Monitors for Environmental Mutagens *Michael D. Shelby*	185
Chapter 17	The Use of Behavioral Techniques in Biological Monitoring Programs *C.L. Mitchell*	191
Part VI	*Critical Issues*	193
Chapter 18	Communications and Dissemination of Information *Robert L. Burgess* and *James Stewart*	195
Chapter 19	Research Needs and Priorities *J.D. Buffington* and *L.W. Little*	197
Chapter 20	Regulatory Control and Public Policy *C. Weber*	201
Chapter 21	Short-Term Organization of Biological Monitoring Methods *Alan W. Maki*	203
Chapter 22	Issues in Long-Term Biological Monitoring *Ken Dickson* and *D.L. Dindal*	213

Index	217
About the Contributors	221
About the Editor	227

Foreword

Biological monitoring for environmental effects is an old concept that has been used by biologists for years. This technical area of biology has, however, only recently been given the emphasis that it deserves. Unlike physical-chemical monitoring methods that require sophisticated instrumentation, computers, and analytical techniques necessitating heavy capital investments, biological monitoring has been given only modest support. Consequently, its potential role in monitoring biological effects has been slowly recognized.

The research findings and recommendations made by the scientists and engineers who participated at this conference/workshop should dispel any doubts about the need to monitor our environment. Biological monitoring is a valuable tool in seeking to understand both reversible and potentially irreversible environmental effects. For responsible authorities in government and private industry not to recognize this would be a serious mistake.

I am hopeful that this book will have an impact upon the future planning of environmental-monitoring programs in the United States and other countries. I also hope it will provide useful guidance in developing programs to evaluate man's impact on his limited natural resources.

Arthur W. Cooper
Head, Forestry Department, N.C.S.U.
President, Ecological Society of America

Preface and Acknowledgments

This book is a result of a conference and workshop that served as a forum for those involved in research, monitoring, legislative development, and the regulatory aspects of biological monitoring of environmental quality. Federal, state, academic, and industrial scientists have exchanged thoughts and formulated recommendations on technical and policy issues in this rapidly growing field.

There is a continuing need to identify the utility of a wide array of bioindicators and biological monitoring techniques for determining both the short- and long-term effects and trends of human activities on our ecosystem and the quality of our environment.

Biological monitoring and physical-chemical monitoring of our environment are at times regarded as separate and competitive. This is unfortunate because they complement each other in almost every way. The response of sensitive indicator organisms to an environmental stress provides an early result. Organisms are uniquely sensitive to multiple environmental stresses operating consecutively or simultaneously. They also integrate the effects of these environmental stresses over time. The ability to selectively accumulate, biomagnify, and show the synergistic effects to exposure from environmental stresses gives bioindicators uniquely useful properties.

The slow recognition of the utility of biological monitoring techniques may be, in part, a result of the difficulty in obtaining quantitative numbers and levels that can be used in a manner similar to that for data obtained by chemical and physical procedures. To some extent, these problems are being solved through the standardization of both laboratory bioassays and in situ biological measurement. Also, increasing sophistication in data collection and interpretation and information on trends in population diversity of ecosystems are providing useful information. Rapid developments and massive amounts of information being gathered through the interdisciplinary activities in federal, state, industry, and academic programs need to be synthesized and documented on a continuing basis.

Aquatic biologists, ecologists, and plant biologists are among those groups of scientists who have been actively developing biological monitoring methods to determine the effects of human activities on biological systems. Communication among these groups has been sporadic at best. One reason for this is the traditional natural inertia between scientific disciplines. Each has its own problems, concerns, and goals that also extend to biological monitoring.

Although the definitions, scope, and roles for biological monitoring have frequently been subjects for discussion within each of the above groups, these groups rarely deliberated on issues of mutual concern. This was one of the primary reasons for planning this book.

Whenever there is a meeting on biological monitoring, the subject of an appropriate definition for biological monitoring is considered. Dr. Cairns offered this useful definition in his provocative presentation: "Biological monitoring is the regular application of biological techniques and methods to determine information about the quality and condition of a biological system." This definition should also be compatible with the purpose that ecologists have applied to biological monitoring, that is, to "monitor the structure and function of natural ecosystems," and to others who use biological indicators for measuring the levels and distribution of environmental contaminants.

Two very real concerns were expressed by several contributors:

1. The lack of clear recognition of biological monitoring in preparing federal and state legislative acts and in environmental regulations. As a result, both funding and staffing are inadequate to do an effective effort in monitoring for effects.
2. The feeling on the part of some scientists that biological monitoring is for applied university-type research or course of instruction.

Fortunately, inroads are being made on both issues. Dr. Buffington elaborated on the efforts of the Council of Environmental Quality to improve overall national environmental-monitoring efforts. It is anticipated that their recommendation will eventually be reflected in a more significant role for biological monitoring in federal research and regulatory control activities.

The rapid progress and growing interest in biological monitoring and the growing need to understand both the subtle and the more obvious effects of stresses we apply to various sensitive species and to biological systems require that universities recognize this area as an integral part of their educational and research activities.

Acknowledgments

Appreciation is extended to the following sponsors of a conference/workshop on biological monitoring for environmental effects, which provided much of the information for this book: Water Resources Research Institute of The University of North Carolina; Research Triangle Institute, North Carolina; Environmental Studies, North Carolina State University; Environmental Studies Council of The University of North Carolina.

The editor wishes to express deep appreciation to the directors and able staff of the Water Resources Research Institute of The University of North Carolina, Raleigh, and especially to Ms. Linda Kiger, administrative officer, and to the efficient, effective assistance received from the head secretary, Ms. Eva McClung.

Part I
Introduction and Overview

The chapters in part I review the philosophical and scientific basis for using biological indicators to monitor the bioeffects of human activities on people and the ecology. One author suggests that biological monitoring is a useful means of measuring people's symbiotic relationship with their natural environment.

The need to reverse a growing public skepticism toward information generated from environmental monitoring programs and the role of biological monitoring in providing more credible bioeffects data is noted. Efforts to upgrade national environmental monitoring programs are described, and alternative futures for biological monitoring are presented. The use of monitoring data in formulating regulatory and legislative environmental policy decisions and in evaluating trends in the quality of our environment is observed. A need to standardize tried and proved biomonitoring methods is discussed.

1 Introduction

Joab Thomas

There are indications of a growing age of skepticism with regard to monitoring the effects of hazardous pollutants in our environment. This may be the result of overreacting on the part of overzealous environmentalists. But for the most part, we have simply been made aware in a number of ways of hazards in the environment. Unfortunately, many people are unable to understand how these hazards affect their lives or what they can do about them. We need to guard against skepticism that develops as the result of misunderstandings or because of inaccurate or misleading information on the effects of these environmental contaminants. A recent newspaper advertisement in Mississippi read, "Azaleas for Sale, Not Known to Cause Cancer."

This was amusing, perhaps, but the depth of meaning went a lot further. We do have this growing skepticism in public attitudes toward alleged effects of environmental monitoring data of various sorts. It is unlikely, however, that there will be much of this skepticism about effects information obtained by biological monitoring methods, since this involves observing the effects of toxic substances on living systems. As we become more sophisticated and able to achieve useful and often quantitative data using biologically sensitive systems, we should have available a network of built-in "smoke detectors" all around the country to give us early warning of pending hazards from environmental pollution.

In North Carolina, for the most part, we still have a relatively clean environment with some time left and the opportunity to guide population and industrial growth so the environment in our state does not become inhospitable to biological systems, including people. We should monitor the effects of our rapid population and industrial growth before it is too late. To do this, it will be necessary to develop and put into effect a credible biological monitoring system.

Hopefully, we can reverse this growing trend toward skepticism in our environmental monitoring programs and develop a sense of reality that will enable us to design and engineer our future rather than to become victims of it.

2 A Review of Environmental Data and Monitoring

John D. Buffington

The President's 1977 Environmental Message[1] provides a useful guide for studies and implementation of changes and improvements in federal programs to maintain the quality of our environment. His message also incorporates innovative administrative initiatives in the environmental area. Some of these have already been brought to satisfactory completion, and others are still in process.

The Council on Environmental Quality (CEQ) received several directives in this message that relate to environmental data and monitoring. Attention was called to large gaps or overlaps in coverage, to poor-quality assurance in data, and to the lack of coordination among laboratories. Also, in many cases, the data obtained suffered from inadequacies in programmatic design. As a result, our environmental monitoring data do not provide federal, regional, or state planning officials with the information they require to assess trends in the quality of the environment or to take legal ameliorative or other needed actions.

In response to the President's request, the Council on Environmental Quality organized an Interagency Task Force on Environmental Data and Monitoring, consisting of five working groups in ecology, air, water, socioeconomics, and land use. Several of the authors of this book are participants in this effort.

One of the problems we immediately encountered is that there is no commonly accepted definition of monitoring. With a great deal of effort, we developed the following: *Environmental monitoring* is the systematic and repetitive collection and analysis of data which can be used (1) to help determine the quality of the environment or condition of natural resources as they are or will be and (2) to help relate environmental quality or natural resources to factors which cause them to change or to effects produced by such changes.

Monitoring is not an end in itself. It is only a means for providing data and analytical information for other functions. For example, three functions support environmental monitoring:

1. Environmental policy and management decisions, including definitions of federal program objectives and priorities and selection of specific regulatory or enforcement actions
2. Identification and definition of environmental problems which are not now recognized or may emerge in the future
3. Evaluation of "progress" in environmental quality as a result of specific federal policies, programs, or actions

Major components of the environmental monitoring process include monitoring design, quality assurance, data management, data analysis, research and development in support of data collection and interpretation, coordination of agency activities, and the review, dissemination, and use of the resulting information.

All you have to do is read the magazines of environmental organizations to realize the frequent dissatisfaction of the private public-interest groups with the way many of our resource decisions are being made. In many cases, it is because the decisions are made in a data-poor environment. The Council on Environmental Quality encourages federal agencies to base their environmental decisions on the soundest possible scientifically valid data.

Major legislation is passed every year assigning tasks to major federal agencies that result in the acquisition of large amounts of environmental data. There is a real need to obtain a clearer picture of what is happening to the nation's environment. It is easier to do this with air quality and water quality information since there are national networks of stations gathering data which are roughly comparable with one another. Many of these data sit in readily available data banks such as STORET and SAROAD. In some cases, the collection method, the frequency of data input, and the institution collecting the data can be made available so that users of these data banks can make independent decisions as to whether the data are valid. For example, CEQ has done precisely this by screening the air and water data through a set of criteria to provide additional quality assurance to erect data sets for analysis in its UPGRADE data analysis systems.[2] Now that this is done, certain questions can be asked. For example, is air quality getting better in five major metropolitan areas or water quality getting worse east of the Mississippi? Are municipal sewage treatment plants helping the nation's water quality situations? Those who have perused the annual reports of the Council on Environmental Quality have seen extensive analyses of such hypotheses for both air and water quality status and trends. These analyses are based on our UPGRADE data sets derived from the Environmental Protection Agency SAROAD and STORET systems.

Those of us who are dealing with *ecological* monitoring, which approximates the biomonitoring topic of this book, have more problems with the definitions of monitoring than those professionally working with air and water quality. For example, although there is much wildlife management information, it is not gathered to assess long-term changes in natural biota. Consequently, little information is available about regional trends and almost none about national trends. Many professionals do not regard such data as constituting ecological monitoring data. Some ecological monitoring programs are not regarded as "ecological" by the performing agencies. Other programs are largely research-oriented.

Each of the working groups is preparing initial analyses of about a dozen issues. These are described in Appendix 2A. These will be synthesized into option papers which will be the basis for implementing changes. In addition, each working group is dealing with several issues proper to its own area.

Several of these issues are being developed by the ecological working group. For example, the need for a program to provide long-term data to monitor the status trends of ecosystems is widely recognized. Currently, the groups dealing with man and the biosphere and the experimental ecological reserves are addressing this issue to some extent. Our study team will examine these programs among others to see precisely what our needs are and identify the best means of satisfying them.

Another topic which is under study is whether a national ecological service is an idea whose time has come. This was first recommended in *The Role of Ecology in the Federal Government.*[3]

A topic of immediate pertinence to this book is impact assessment monitoring. Several recent symposia, including one sponsored by CEQ at the 1976 American Institute of Biological Sciences (AIBS) meeting, have already addressed this topic. There appears to be consensus that improvements are possible. We will suggest a program of specific recommendations.

In inventorying ecological monitoring programs, we found that many agencies do not have monitoring as a discrete budget item. Also much of what is going on in the monitoring area is taking place in regional organizations of federal agencies. These regional organizations themselves do not have monitoring broken out in their budget. In some cases, the regional budgets of agencies are lost at the federal level. We dealt as best we could with these problems and assembled an inventory which describes as a first cut the ecological monitoring taking place in the federal system. Unfortunately, it is very difficult even to precisely identify the resources being applied in this area.

Preliminary results of our effort suggest that ecological monitoring is where air and water monitoring were perhaps twenty years ago. This accounts for the large research component and the lack of centralized focus. Moreover, there is no overriding legislative mandate such as there is for the physical media. However, I anticipate that the progressive suggestions and contributions made in this book will help to evolve and systemize our thinking on this subject of biological monitoring.

Notes

1. The President's Environmental Program, Washington: Government Printing Office, 1977.

2. The *U*ser *P*rompted *GRA*phic *D*ata *E*valuation system was developed by CEQ with Environmental Protection Agency and Department of Energy cosponsorship and the participation of other agencies. It is completely user-oriented since it is English-prompted and requires no specialized knowledge of computer language. It permits rapid graphic and statistical analysis of environmental data.

3. Council on Environmental Quality and Federal Council for Science and Technology, *The Role of Ecology in the Federal Government*, December 1974.

Appendix 2A: Task Force Multidisciplinary Topics

1. *Coordination*
   What steps, if any, should be taken to improve the interagency coordination of environmental data and monitoring programs? How should the recommendations of the President's Reorganization Project be implanted in the area of environmental monitoring?
2. *Information Needs and User Feedback*
   What steps, if any, should be taken to improve the identification of priority data and information needs to provide better guidance to monitoring programs and ensure better feedback from federal and nonfederal users? What kind of user needs inventory can and should be developed by the task force?
3. *Monitoring Design*
   What steps, if any, should be taken to improve the design, technical plans, and planning mechanisms for the data collection phase of monitoring programs, to improve the usefulness of their data and increase cost efficiency?
4. *Quality Assurance*
   What steps, if any, should be taken to identify and improve the accuracy, reliability, and scientific validity of the data collection?
5. *Data Management*
   What steps, if any, should be taken to improve the usefulness of data systems for analysis and interpretation, increase the comparability of data for efficient cross-disciplinary analyses, and increase the efficient accessibility of the data to a wider user community to foster multiple use of data and independent analyses? What would be the impact of a national environmental data system such as that proposed in the 1972 Dingell Bill?
6. *Evaluation and Reporting*
   What steps, if any, should be taken to improve the relative emphasis on and mechanisms for the analysis, evaluation, and dissemination of environmental data?
7. *Nonfederal Data Sources and Needs*
   What steps, if any, should be taken to better meet the most important needs of the public, state and local agencies, academics, industry, and so on? What in the private sector should be stored and distributed by the federal government?
8. *Priority Research Needs*
   What are the most critical R&D needs for monitoring methods, environmental assessment to interpret monitoring data, and understanding relevant physical and chemical processes?

9. *Resource Allocations*
   What are the present patterns of federal resource allocation to monitoring and related activities? Are these patterns reasonable? Can we identify some needed changes now? Can we suggest better mechanisms to assist in planning these allocations in the future?
10. *Anticipatory and Reference Monitoring*
    What steps, if any, should be taken to improve monitoring to better identify emerging or unforeseen environmental problems and trends?
11. *Global and Regional Monitoring*
    What should be done to monitor wide-scale environmental changes in areas such as the northern Great Plains, the global ocean, and the global atmosphere?
12. *Program Evaluation Monitoring*
    What steps, if any, should be taken to improve monitoring to evaluate the effectiveness of environmental management and pollution control policies, programs, and actions?
13. *Source Monitoring*
    What steps, if any, should be taken to improve our national data on pollutant emissions/discharges, source compliance, and related topics?

3 Scenarios on Alternative Futures for Biological Monitoring, 1978-1985

John Cairns, Jr.

The agricultural revolution occurred because the unmanaged environment would not produce sufficient food to keep our society reliably and adequately fed. The "biological monitoring revolution" is now occurring because the unmanaged environment is unable to assimilate social wastes without being harmed. For those who desire both the advantages of a technological society and the amenities of natural systems (including our dependence on them as a life support system), it is mandatory that the interface between the two systems be as harmonious as possible.

There are many similarities between ecosystems and individual organisms which deserve attention: (1) both consist of interlocking-interactive cause-effect pathways; (2) both collapse if energy-nutrient inputs cease and change if either is altered; (3) both have evolved homeostatic mechanisms to compensate for natural perturbations. Although these comparisons may seem obvious to the point of triviality, present regulation of waste discharges is not based primarily on these factors. Pollution control legislation must recognize that drainage basins are systems that require feedback to maintain homeostasis. The process of developing feedback is called biological monitoring, which should always be accompanied by chemical-physical monitoring.

It is impossible to properly manage any system without a continual feedback of information about its condition. Since all the biological processes and some of the chemical-physical processes in treatment plants are similar or identical to those that occur in natural systems, waste treatment facilities may be regarded as river or lake extenders and therefore as an integral part of the water ecosystem. Instead of being designed to fit artificial arbitrary "pipe standards," waste treatment facilities should be designed so that their products are compatible with the receiving system's assimilative capacity. A second important point is that treatment facilities and their discharges should be considered collectively in terms of their interactions, systems effects, and so on, rather than as separate, unrelated entities. The most likely future scenario couples biological monitoring with ecosystem management.

Environmental Management Revolution

Revolution is perhaps a dramatic word for something that occurs gradually. However, we speak of the agricultural revolution which certainly did not occur overnight. That particular revolution occurred because the unmanaged environment was not capable of producing sufficient food to keep society reliably and adequately fed in terms of the aspirations of that time. In short, merely harvesting or using the dietary benefits to be derived from natural systems without contributing anything toward their management and continuity was no longer adequate. Almost since the beginning of the industrial revolution we have been viewing the environment in much the same way as the hunters and gatherers who preceded the farmers into the environment without regard for its ability to assimilate these. And the objectional consequences are everywhere visible. Even the Best Applicable Technology (BAT) is a sophisticated version of the "ignore the environment" theme. Essentially, what it says indirectly is this: We will use our best technology to resolve waste treatment problems, and the environment will just have to cope with the result. Failure to determine the response of the environment to the waste is justified as a combination of administrative convenience, simplicity of tests, and economics. However, if the ecosystem on which we depend for both necessities and amenities of life is destroyed in this process, it will be because we failed to alert the management system to the apparent limited capabilities of aministrators who cannot see beyond simplistic solutions to complex problems, economists who forget to put environmental costs and benefits into their equations, and technicians who would rather do a simple test than a difficult one.

No industrial quality control person would think of depending on a system which did not involve (1) placement of sensors at critical spots throughout the system, (2) rapid generation of information about the quality of the system, and (3) development of an organization capable of taking corrective action when things go wrong. Translated to environmental problem solving, this means we immediately couple environmental management and biological monitoring accompanied by chemical-physical monitoring. In other words, it is arrogant to assume that we can manage a system without a continual feedback of information about its condition. This feedback of information about the biological portion of the system is called biological monitoring. The term *biological monitoring* has been grossly misused in recent years. It is being used to describe assessments that were carried out only once or erratically. *Biological monitoring* is the regular application of biological assessment techniques and methods to determine information about the quality and condition of a biological system.

Justification of Costs

Sanitary engineers find it comparatively easy to get out the first 90 or 95 percent of the contaminating material (with some notable exceptions). It is several to many times more costly to get out the remaining 5 percent. It makes sense to cope with the major portion of the contaminating material when it is in its most concentrated and most easily treated form before it is discharged into natural ecosystems. This is the cheapest and most efficient way since one gets the greatest results for the least money. Whenever it is possible to do so without degrading the biological integrity of the receiving system (that is, displacing it in either structural or functional characteristics), the most expensive part of the treatment may be carried out in natural systems. The economic benefits to industries permitted to do this are equivalent to the cost of (1) the additional capital facilities for complete treatment, (2) the operating costs including personnel, and (3) the interest on bonds required to finance complete treatment. If charges for even a fraction of these were made, financial resources would be provided for the ecological research, monitoring, and personnel training that are so badly needed. Properly generated data would enhance our understanding of some fundamental ecological processes and increase demand for fisheries personnel and other types of biologists. There are the usual dangers that regulators will become advocates of this group they are charged with regulating and that to meet the increased demand, positions would be filled with improperly trained personnel.

There are few precedents for environmental-use charges, but it is quite likely that frequently they would provide an incentive for additional waste treatment in manmade facilities. Industry would benefit because more flexibility and options would be provided than with the present system, which does not adequately recognize that some ecosystems are more vulnerable than others and tends to set one standard for the entire country. This system penalizes industry where excessive treatment (for which there are no biological benefits) is required and hurts ecosystems when the treatment required is inadequate [1].

When Should Biological Monitoring Be Used?

Biological monitoring should be instituted whenever a waste discharge or other potential environmental stress has a significant possibility of harming the receiving system. During an environmental hazard evaluation procedure, increasingly accurate estimates are obtained about concentrations of the chemical (or other stress) that do not cause adverse biological effects and the environmental concentrations that will result from production and use of the chemical. The process of hazard evaluation may be depicted graphically. In figure 3-1 the "no adverse biological effect" level and the concentration that will result from introducing the chemical into the environment are well apart [5]. In fact, only estimates of these concentrations are known and are indicated by dotted lines which envelope the solid concentration lines. This is because tier I testing consists of comparatively crude short-term tests. Tier II testing is more sophisticated and expensive (for example, continuous flow instead of batch), and tier III is even more so. Frequently, tier I testing will improve the estimates sufficiently

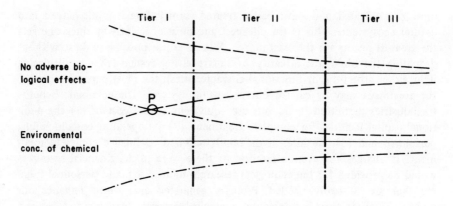

Figure 3-1. Conditions Where Short-Term Testing Show Hazards to Be Low

that one will be able, at decision point P, to determine that the concentrations are indeed different. Thus, there may be justification for terminating testing in tier I and concluding, *at a certain risk level,* that introduction of the chemical will not cause an environmental hazard. One should remember that risk cannot be reduced to zero. Both ecology and economics have externalities!

In figure 3-2 the two concentrations are closer together. Now, testing must be carried through tier III before the same statement can be made at a comparable risk level. Figure 3-3 depicts the case where testing up to the same point leads one to conclude that the chemical would cause an environmental hazard because the environmental concentration would be greater than the "no adverse biological effects" concentration.

This simple concept should be well understood because it is the basis of all hazard evaluation procedures. It also demonstrates that biological data alone are inadequate for hazard assessment. Of course, biologists have been saying for years that chemical data alone are also inadequate. Fisheries scientists and other biologists must develop an effective working relationship with production engineers, chemists, chemical engineers, marketing personnel, and a variety of other disciplines and activities in order to produce an adequate hazard evaluation. Since the various types of evidence in the successive tiers must be "orchestrated," biologists must learn to exchange information continually with other groups and *to react* (including redesigning experiments) to the information they provide [2].

In short, biological monitoring should be accompanied by chemical-physical monitoring and should be carried out whenever there is a high probability of risk to the ecosystem receiving the waste or other pollutional stress. The degree of risk is determined by the relationship between environmental concentration of the chemical and the concentration which produces no adverse biological effects.

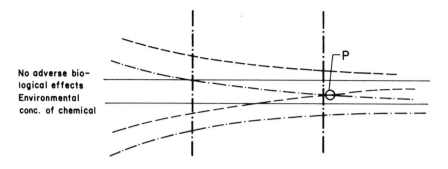

Figure 3-2. Conditions Where Extensive Testing Needed to Show Hazards to Be Low

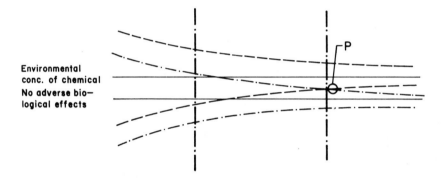

Figure 3-3. Conditions Where Extensive Testing Is Needed to Show Probable Bioeffects

Options in Regulating Waste Discharges

Although there appears to be fairly widespread dissatisfaction with the present systems of regulating waste discharges, there is no consensus for alternative approaches. Three alternatives seem to be available: (1) improvement of current practices by using all available scientific information and disseminating this information through such books as the EPA's *Water Quality Criteria of 1972* and *Principles for Evaluating Chemicals in the Environment 1975* [6]; (2) use of maximum feasible treatment with available technology on each waste discharge with the assumption that technology will improve substantially and therefore environmental conditions will improve; and (3) use of biomonitoring techniques to fully permit nondegrading utilization of the receiving of assimilative capacity of ecosystems for waste discharges and at the same time protect them from deleterious effects. Alternative 2 is espoused by a variety of groups, but it is unsatisfactory to others for a number of reasons. There are two prominent reasons: (1) Reconstruction of the treatment system might be necessary with each major technological advance, and there is no way of predicting when all this would end or how frequently changes would be necessary; (2) there might be no measurable or demonstrable environmental benefits resulting from the improved waste treatment even though it might be considerably more expensive than the systems replaced.

Objections to biological monitoring are that there is no strong body of evidence indicating that it will work as it is supposed to nor is there a sizable body of evidence on cost. If the cost is high, small industries might be priced out; and in any case, the monitoring must still be accompanied by a well-functioning waste treatment system. There will also be difficulties in formulating appropriate legislation.

Neither system alternative to the one presently used for regulating waste discharges is likely to replace it without massive evidence of superior effectiveness at acceptable cost. This evidence of superiority should be strong in three principal areas: (1) scientific justifiability, (2) operational reliability, and (3) cost-effectiveness. Before areas 2 and 3 are addressed, area 1 must be supported to the satisfaction of a substantial segment of the academic community. Sufficient evidence must be gathered for the academic community to make a tentative judgment on the efficacy of biological monitoring and identify questions which must be answered before substantial confidence can be placed in biological monitoring. A thorough evaluation should be made of both the conceptual soundness and the operational capabilities of various biological monitoring methods. Although some indirect evidence will be available for the cost-effectiveness of these methods, it would be presumptuous to attempt a detailed evaluation at such an early developmental stage for the field. However, it appears likely that the use of minicomputers and other technological innovations will result in a substantial saving, as they have in other areas of applied science.

Many professions, such as engineering, regard the production of standard methods as a very important professional responsibility. Perhaps the major reason has been the need for standard methods to assess parameters deemed important for public safety and health so that regulatory agencies, manufacturers, and others might carry out tests that are comparable. Despite the common acceptance of the need for the development of standard methods in many disciplines, this is not true of biology [2].

Until recently the need for standard biological methods has not been acute except in certain specific areas such as public health, food production, and the like. Regulatory agency interest in ecosystem structure and function (at least in terms of specific assessments) was not prominent until the post-Earth Day period when environmental protection became a major responsibility.

Since one of the best measures of the health of an ecosystem is the condition of both the individual species and communities of indigenous biota, development of standard methods for sampling and assessing conditions has become mandatory. In this regard society, particularly through the courts, asks two basic questions of members of a profession: What parameters or performance characteristics are most important and therefore worth measuring? and What are the tried-and-true, generally accepted methods for making these measurements? This means communicating not the latest and most exciting developments in the field but rather the "old workhorse" methods even though these may have lost some of their luster as research tools because some deficiencies have become apparent through continued use. This is precisely why they become attractive as standard methods even though they may have lost some of or all their appeal to research investigators.

The requirements of a standard method are many, but a few of the most important are: (1) It can be used by practitioners other than research investigators with assurance of reliable, reproducible results. (2) It has been used sufficiently broadly that its major deficiencies are known and documented and the best ways of avoiding these deficiencies have been identified. (3) It is known to and used by a substantial segment of the profession and has proved utility. (4) It has gone through a generally accepted process of evaluation before becoming a standard method.

There are several ways to contribute to the development of standard methods. A description follows of the American Society for Testing and Materials (ASTM) subcommittee on biological monitoring, which is part of Committee D19 on water. There are other fine standard-setting groups (for example, the American Public Health Association), but space does not permit a description of each. An outline of the activities of the ASTM group concerned with biological monitoring follows but is not accompanied by detailed descriptions of all the specific methods being developed under each category:

Subcommittee D19.20—Biological Monitoring
    Coordinating committee for D19.21, D19.22, D19.23, D19.24, D19.25

Subcommittee D19.21—Effluent Testing
    Representative Task Force Activities
        Acute toxicity tests for effluents
        Algal bioassays
        Flavor impairment
        Inhibitory toxicity

Subcommittee D19.22—Metabolic and Physiological Activities of Organisms
    Representative Task Force Activities
        ATP method
        Oxygen uptake

Iron bacteria
Sulfate bacteria
Microscopic matter

Subcommittee D19.23—Biological Field Methods
Representative Task Force Activities
Phytoplankton
Macrophyton
Microinvertebrates
Fish

Subcommittee D19.24—Bacteriological Indicators
Representative Task Force Activities
Coliforms
Fecal coliform
Fecal streptococci
Pseudomonas aeroginosa
Vibrio parahaemolyticus
Clostridium perfringens
Aeromonas hydrophila
Coagulase positive staphylococci
Candida albicans

Subcommittee D19.25—Statistical Methods
Just formed. The initial activity is a symposium "Quantitative and Statistical Analyses of Biological Data in Water Pollution Assessment." The proceedings will be published by ASTM and will serve as a base for standard method development.

For those who wish to become involved, it is possible to volunteer contributions by being part of a task force to see a specific method through the ASTM system, by being part of one of the categories which evaluates a spectrum of methods, or by helping at a policy-setting level on the coordinating committee. It is essential that people who get involved in the latter way have experience in the other activities. It is not necessary to be an ASTM member in order to produce a method, but it is essential if one wishes to vote on the efficacy of a method. Individual membership is not exorbitant (some academic institutions are already institutional members and may send faculty without additional charge) and includes receiving a free copy of one of the standards books (for example, the one on water) produced annually. Possibly, the best way to decide whether this is worthwhile is to attend one of the meetings or become involved in a task force charged with developing a specific method.

Development of standard methods is not an exciting process and often can be extremely frustrating because it involves consensus among diverse points of view. It is a badly neglected professional responsibility which biologists must address.

Enhancing the Acceptability of Biological Monitoring

There are many questions regarding the conceptual soundness of biological monitoring methods that must be answered more definitively before these methods will be accepted by the academic community, potential industrial users, and regulatory agencies. For an in-plant monitoring system, the most important questions are probably the following:

1. Will the system detect spills of lethal materials before they reach the receiving waters?
2. If only one organism is used as a sensor (for example, the bluegill sunfish), will this organism be so much more tolerant to the particular toxicant in question that the chemical will pass undetected and harm other members of the aquatic community in the receiving system (for example, algae and invertebrates)?
3. Is it possible to monitor chemical-physical parameters and achieve the same results at lower cost and greater efficiency?
4. Since the biological response alone will not identify the particular toxicant causing the response but only indicate that some deleterious material is present, is it possible to couple a chemical-physical monitoring system with a biological monitoring system that will expedite the identification of the particular deleterious component causing the warning response?
5. Will a false signal cause an expensive shutdown of the plant or an undue expenditure of time and effort by the waste control personnel?
6. Should an organism indigenous to each receiving system be used, which would require a long site-specific developmental period for each new drainage basin, or can some all-purpose organism, such as the bluegill, be used for all types of systems (or perhaps one organism for a warm-water and one for a cold-water system)?
7. Is it possible to use in-plant biological monitoring systems to detect the presence of spills of materials having acute lethality or long-term effects or only the former?
8. Are the in-plant monitoring systems only for very large industries with sizable waste control staffs, or is it possible to develop compact, miniaturized, reliable in-plant monitoring systems that can be used by persons inexperienced in monitoring without undue expenditure of time and so forth?

Questions such as these related to the utility and scientific justifiability of monitoring will undoubtedly be asked by persons representing regulatory agencies and the industrial point of view. These eight questions were raised primarily because they are important in the acceptance of biological monitoring, and they indicate how far we must yet go in developing the methods. For example, it took nine years before the apparatus and methods for in-plant biological monitoring were ready for use on a trial basis in an industry [8].

During this developmental period, the apparatus has gradually evolved from strip charts and other visually examined data recording systems to the present computer-interfaced, automated data recording system. In addition, a variety of responses were examined (for example, the coughing response, [7]) which are presently being used by the Environmental Protection Agency [3]. In the course of this investigation, it became apparent that any characteristic of an organism's normal life which was evident continually (for example, respiratory signals of fish) could be used effectively in a biological monitoring system. A key factor in successful use was the ability to quickly and reliably detect deviations from the normal condition. Although some attention has been given to the development of statistical methods for this purpose [4], much more attention must be given to this critical area.[1]

No system can be adequately managed without feedback of information about its condition! There is widespread recognition of this simple fact: biological monitoring and environmental management will continue to take the back seat to such inappropriate measures as Best Applicable Technology. Techniques are already available for adequate measures, but we need to develop many more. Universities should realize that persons developing such techniques must make efforts to develop sound biological monitoring methodologies and to convert to standard methods those methods which are already available and generally known to practitioners in the field. Finally, they must convince colleagues in the academic community that it is not a disgrace to work with practical problems and that they are not void of conceptual merit.

My original assignment was to cover the years 1978 to 1985. There are only two possible scenarios. In the first one, the persons now engaged in the development of biological monitoring methodologies will fail to convince others of their utility, and alternative criteria will be used for environmental protection. It would be foolish to ignore the fact that biological monitoring is accepted by only a minority of biologists and engineers and is hardly understood by politicians, policymakers, and heads of industries. The second possibility is that in the period of 1978 to 1985, biological monitoring will become a routine part of waste and environmental management. There are some hopeful signs. First, biological monitoring is now being developed in every major industrialized society and some developing countries. Second, symposia workshops such as the one that prompted this book are further fostering the development of the field by bringing together those who have a deep interest in the subject and a wish to share information. Third, many federal and some state laws now mention biological monitoring although they do not state precisely what form it should take.

It would be a disservice to say that the future of biological monitoring is ensured despite the clear, almost compelling need for it. The alternative scenarios of success and abject failure for the field will depend on our ability to communicate the desirability of biological monitoring and the soundness of the methodologies and techniques between now and 1985.

Note

1. Quoted with permission from J. Cairns, Jr., K.L. Dickson, and G.F. Westlake, eds. *Biological Monitoring of Water and Effluent Quality*, STP 607 (Philadelphia: American Society for Testing and Materials, 1977), pp. 235-238.

References

[1] Cairns, J., Jr. Aquatic ecosystem assimilative capacity. *Fisheries* 2(2):5-7, 24 (1977).
[2] Cairns, J., Jr. Hazard evaluation. *Fisheries* 3(2):2-4 (1978).
[3] Drummond, R.A., Olson, G.F., and Batterman, A.R. Cough response and uptake of mercury by brook trout, *Salvelinus fortinalis,* exposed to mercuric compounds at different hydrogen-ion concentrations. *Transactions of the American Fisheries Society* 103(2):244-249 (1974).
[4] Hall, J.W.; Arnold, J.C.; Waller, W.T.; and Cairns, J., Jr. A Procedure for the detection of pollution by fish movements. *Biometrics* 31(1): 11-18 (1975).
[5] Mill, T.; Smith, J.H.; Mabey, W.; Holt, B.; Bohonos, N.; Lee, S.S.; Bomberger, D.; and Chou, P.W. Environmental exposure assessment using laboratory measurements of environmental processes. *Proceedings of the Ecosystem Symposium*, Corvallis, Ore. 1977, pp. 1-21.
[6] National Academy of Sciences (NAS). *Principles for Evaluating Chemicals in the Environment.* Washington: NAS, 1975.
[7] Sparks, R.E., Cairns, J. Jr., McNabb, R.A. and Suter, II G. Monitoring zinc concentrations in water using respiratory response of bluegills (*Lepomis macrochirus Rafinesque*). 40(3):361-369 (1972).
[8] Westlake, G.F. and van der Schalie, W.H. Evaluation of an automated biological monitoring system at an industrial site. in J. Cairns, Jr., K.L. Dickson, and G.F. Westlake, eds. *Biological Monitoring of Water and Effluent Quality*, pp. 30-37. Philadelphia: American Society for Testing and Materials (1977).

Part II
Monitoring Bioeffects of Water Pollution

Authors in part II describe government, industrial, and academic programs for developing and applying biomonitoring methods to assess the effects of water pollution. Federal water and state legislative mandates requiring bioeffects information on receiving waters are described. Concern is expressed for inadequate training programs in universities and staffing of regional laboratories to be responsive to regulatory programs and the need for developing improved biological monitoring methodology. Research and development activities underway and recommended for developing standard biomonitoring techniques and procedures for obtaining scientifically credible information are also described.

4 Federal and State Biomonitoring Programs

Cornelius I. Weber

One of the primary goals of the Federal Water Pollution Control Act (FWPCA), as amended in 1972 and 1977, is to restore and maintain the biological integrity of the nation's waters. Although the *biological* integrity of water is not explicitly defined in the act, frequent mention is made of the protection and propagation of fish, shellfish, and wildlife and the effects of pollutants on the diversity, productivity, and stability of communities of indigenous aquatic organisms. Emphasis is also placed on determining the biological properties (toxicity) of effluents and the effects of effluents on the biota of receiving waters. The definition of biomonitoring in the law is very broad and includes the determination of the effects of pollutants on (all) aquatic life.

The Environmental Protection Agency (EPA) has identified three types of water monitoring: compliance (discharge permit) monitoring, ambient (water quality) monitoring, and intensive surveys. Agency guidance for biomonitoring programs has been issued in two reports, *Model State Water Monitoring Program* (EPA 1975) and *Basic Water Monitoring Program* (EPA 1977).

The responsibility for the ambient monitoring program has been delegated largely to the states, through a federal grant program. The internal EPA monitoring program is carried out through the Regional Surveillance and Analysis Divisions and is limited largely to compliance monitoring and intensive surveys. The EPA also has initiated a limited ambient water monitoring program called the National Water Pollution Surveillance System.

Extensive biomonitoring programs have been, and are now being, carried out by the EPA and other federal, state, and private agencies. The FWPCA places strong emphasis on the restoration and maintenance of the biological integrity of the nation's waters and makes frequent mention of the protection and propagation of the indigenous communities of aquatic organisms. To be effective, biomonitoring programs must include measurements of the toxicity and/or biostimulatory properties of effluents and the effects of effluents on aquatic life in receiving waters (ecosystems). All communities of indigenous aquatic organisms should be sampled, where appropriate, including the plankton, periphyton, macrophyton, macroinvertebrates, and fish.

The properties of communities of aquatic organisms utilized to examine the biological integrity of surface waters and to describe the effects of pollutants on the aquatic organisms in receiving waters are included in the following three basic categories: (1) Standing crop—the numbers and biomass (size, weight, and so on) of organisms present per unit area or volume (population density),

(2) community structure—the kinds (species) of organisms present and their relative abundance, and (3) community metabolism and condition—the rate of physiological processes (such as photosynthesis, respiration, and nitrogen fixation), the bioaccumulation of toxic substances, and the occurrence of pathological conditions.

Where appropriate, representative species are used in controlled laboratory, treatment plant, and field studies of the toxic and/or biostimulatory properties of effluents.

The EPA has prepared guidelines for biomonitoring programs and has published methodology for biological field and laboratory studies in a manual entitled *Biological Field and Laboratory Methods for Measuring the Quality of Surface Waters and Effluents* (Weber 1973a). A national standing committee of senior EPA biologists provides assistance in screening, selecting, and describing available methodology to be incorporated in the manual and in identifying new methodology needs.

Introduction

Objectives of Federal Water Pollution Control Legislation

One of the principal objectives of the 92d Congress in passing the 1972 amendments to the Federal Water Pollution Control Act (FWPCA) (Public Law 92-500) was to restore and maintain the biological integrity of the nation's waters and to achieve, by July 1, 1983, wherever attainable, a quality of water that provides for the protection and propagation of aquatic life. Recognizing the interdependence of human health and welfare and aquatic life, Congress included in this legislation the authorization and/or directives for the U.S. Environmental Protection Agency and the state programs to conduct comprehensive biological monitoring programs. Section 502(15) of the act defined biological monitoring as "the determination of the effects on *aquatic life*, including the accummulation of pollutants in tissue, in receiving water due to the discharge of pollutants (A) by techniques and procedures including sampling of organisms representative of appropriate levels of the food chain appropriate to the volume and the physical, chemical and biological characteristics of the effluent, and (B) at appropriate frequencies and locations." Other sections of this act, and its successor (the Clean Water Act of 1977), refer to measurement of the biological properties of effluents, the effects of toxic and heated effluents on the aquatic life in receiving waters, and the trophic status of recreational lakes.

The development of most of the technical details of the design of the biomonitoring program required to attain the goals of Public Laws 92-500 and 95-217 was left to the EPA. However, the legislation in defining the scope of the program makes frequent use of terminology such as *diversity, stability, and productivity* which relates to specific biomonitoring parameters. The assurance that aquatic life is protected and is propagating, as required by the act, can be achieved only through a coordinated, comprehensive national biomonitoring program that is adequately staffed and includes various federal, state, and local government agencies and private organizations.

Legislative Mandate for Biological Monitoring

The legislative mandate for the collection of biological data by the Environmental Protection Agency and other federal, state, and private agencies is either clearly stated or implied in at least nineteen sections of the federal Pollution Control Act Amendments of 1972 and 1977. Some of the more prominent examples are listed below [see section 502(15) for the definition of biological monitoring] :

Sec. 101 (a)		The objective of this Act is to restore and maintain the chemical, physical, and *biological* integrity of the Nation's waters.
	(2)	it is the national goal that wherever attainable, an interim goal of water quality which *provides for the protection and propagation of fish, shellfish, and wildlife*...
Sec. 104 (b)		... the Administrator is authorized to ... collect and disseminate ... basic data on chemical, physical, and *biological effects* of varying water quality...
	(6)	
	(d)	In carrying out the provisions of this section the Administrator shall develop ...
	(2)	Improved methods and procedures to identify and measure the *effects* of pollutants...
Sec. 105 (d)		In carrying out the provisions of this section, the Administrator shall conduct, on a priority basis, an accelerated effort to develop, refine, and achieve practical application of:
	(3)	improved methods and procedures to identify and measure the effects of pollutants on the chemical, physical, and *biological integrity* of water, including those pollutants created by new technological developments.
Sec. 106 (e)		[Refers to the state monitoring programs which must provide for...]
	(1)	the establishment and operation of appropriate devices, methods, systems, and procedures necessary to monitor,

	and to compile and analyze data on (including classification according to *eutrophic condition*), the quality of navigable waters... *including biological monitoring*... [Also see sec. 305(b).]
Sec. 302(a)	Whenever...discharges of pollutants...would interfere with the attainment or maintenance of that water quality...which shall assure...the *protection and propagation of a balanced population of shellfish, fish, and wildlife*...effluent limitations...shall be established....
Sec. 303(d)(1)(b)	Each state shall identify those waters or parts thereof within its boundries for which controls on thermal discharges...are not stringent enough to *assure protection and propagation of a balanced indigenous population of shellfish, fish, and wildlife.*
Sec. 304(g)	The Administrator shall...promulgate guidelines establishing test procedures for the (biological) analysis of pollutants....
(h)	The Administrator shall...promulgate guidelines for ... the acquisition of information from owners and operators of point sources of discharges...which shall include: (A) (biological) monitoring requirements...
Sec. 305(b)	Each state shall prepare and submit to the Administrator ...each year...a report which shall include...(B) An analysis of the extent to which the navigable waters... provide for the *protection and propagation of shellfish, fish and wildlife, and allow* recreational activities in and on the water, (C) An analysis of the extent to which the elimination of the discharge of pollutants and a level of water quality which provides for the *protection and propagation of a balanced population of shellfish, fish and wildlife*... have been or will be achieved.
Sec. 308(a)(3)	(A) The Administrator shall require the owner or operator of any point source to.... (iii) install, use, and maintain such monitoring equipment or methods...including where appropriate, *biological monitoring* methods...
Sec. 311	The Administrator shall develop...regulations...(pertaining to) hazardous substances...(which) present an imminent and substantial danger to...*fish, shellfish, wildlife*....

Federal and State Biomonitoring Programs

Sec. 314(a) Each state shall prepare or establish, and submit to the Administrator for his approval
(1) an identification and classification according to *eutrophic condition* of all publicly owned fresh water lakes in such State...

Sec. 316(a) ...The Administrator may impose an effluent limitation... that will assure the *protection and propagation* of a *balanced, indigenous population of shellfish, fish, and wildlife* in and on the body of water (into which the discharge is made)

Sec. 403(c)(1) The Administrator shall... promulgate guidelines for the determining of the degradation of the waters of the territorial seas, the contiguous zone, and the oceans, which shall include:
(A) the *effect of disposal of pollutants on*... but not limited to *plankton, fish, shellfish, wildlife*...
(B) the effect of disposal of pollutants on marine life including the transfer, concentration, and dispersal of pollutants or their byproducts through *biological*, physical, and chemical processes; *changes in marine ecosystem diversity, productivity, and stability; and species and community population* changes...

Sec. 502(15) *The term "biological monitoring" shall mean the determination of the effects on aquatic life, including accumulation of pollutants in tissue, in receiving waters due to the discharge of pollutants (A) by techniques and procedures, including sampling of organisms representative of appropriate levels of the food chain appropriate to the volume and the physical, chemical, and biological characteristics of the effluent, and (B) at appropriate frequencies and locations.*

Objectives of a Biomonitoring Program

The threat to human health and welfare posed by the pollution of surface waters has two fundamental aspects: the *direct effects* on human health through contaminated water supplies and food and the *indirect effects* resulting from the impact of pollution on the quantity and quality of aquatic organisms used for human food, the use of water for recreation, the aesthetic quality of the environment, and the integrity of the biosphere (figure 4-1).

The objectives of a biomonitoring program based on Public Law 95-217 are to determine:

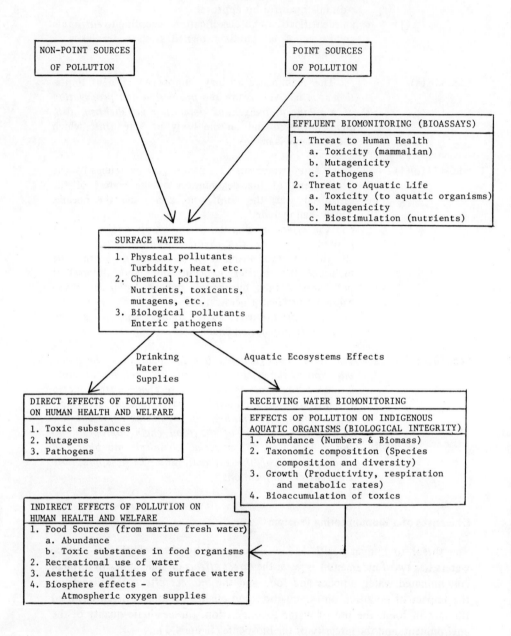

Figure 4-1. Biological Effects of Pollution on Human Health and Aquatic Life

Federal and State Biomonitoring Programs

1. If toxic substances and pathogens are being discharged to surface (and ground) waters
2. The dispersion and persistency of pesticides, toxic metals, and other toxicants in water and aquatic life [Sec. 104(1), (2); 304(a)(1); 403(c)(1)(A)]
3. If aquatic life has indeed been "protected" and is "propagating" [Sec. 101(a)(2)]
4. Long-term trends in the "diversity, productivity and stability" of communities of indigenous aquatic organisms [Sec. 304(a)(1)]
5. The "trophic status" of surface waters and long-term trends in eutrophication of surface waters [Sec. 106(e)(1); 314(a)]
6. Undesirable effects of nutrients in surface waters [Sec. 104 (h)(A)]
7. If effluent guidelines and permit limitations (NPDES) are adequate to protect aquatic life in receiving waters [Sec. 308(a); 309(a)]
8. The effects of thermal discharges [Sec. 303(d); 316]
9. The effects of spills of oil and other hazardous materials (Sec. 311)
10. The success of the program for the "rehabilitation and environmental repair" of Lake Erie [Sec. 108(d)(1)]
11. The present and projected quality of the waters of the Great Lakes [Sec. 104(f)]
12. The efficacy of domestic waste treatment plants
13. If the goals of the act are being met
14. If program resources are properly allocated

Biological Integrity

From the wording of the various sections of Public Law 95-217 relating to biological monitoring, and especially section 502(15), it is abundantly clear that it was the intent of Congress, in using the terms *biological integrity, aquatic life, plankton, shellfish, fish, and wildlife, ecosystem (population diversity, stability, and productivity)*, and so on to include all communities (types) of aquatic life, freshwater and marine.

The biological integrity of surface waters is related to the following basic questions: Is the water free of hazardous substances? Are the expected kinds of aquatic organisms present in the expected numbers, carrying out life functions at normal rates, free of toxins and pollutant-related disease?

The parameters that must be measured in a biomonitoring program designed to satisfy the requirements of Public Law 95-217 fall into two major groups (see tables 4-1 and 4-2)

1. Biological Properties of Pollutants
   A. Toxicity to human and aquatic life
      (1) Specific pollutants
      (2) Complex (mixed) effluents

Table 4-1
Properties of Indigenous Communities of Aquatic Organisms Used in Determining the Biological Integrity of Surface Waters

Parameters	Phyto plankton	Zoo- plankton	Peri- phyton	Macro- phyton	Macro- invert	Fish
Standing Crop						
1. Count	X	X	X	X	X	X
2. Volume	X	X	X		X	
3. Wet weight	X	X	X	X	X	X
4. Dry weight	X	X	X	X	X	
5. Ashfree weight	X	X	X	X	X	
6. DNA content	X					
7. ATP content	X	X	X			
8. Chlorophyll *a* content	X		X	X		
Taxonomic Composition						
1. Species identification	X	X	X	X	X	X
Indicator species	X	X	X	X	X	X
Number of individual species	X	X	X	X	X	X
Total number of species	X	X	X	X	X	X
Diversity index	X	X	X	X	X	X
2. Pigment composition						
Biomass/Chlorophyl *a*	X		X			
Chlorophyl *a*/Chlorophyl *b*	X		X			
Chlorophyl *a*/Chlorophyl *c*	X		X			
Pheophytin content	X		X			
3. Nitrogen (N_2) fixation	X		X			
Metabolic Activity or Condition						
1. *Primary productivity*						
Carbon-14 uptake	X		X	X		
Oxygen evolution	X		X	X		
2. *Respiration rate*						
Plankton dark-bottle O_2 uptake	X					
Electron transport	X					
Benthic respirometer O_2 uptake			X	X	X	
3. *Nitrogen (N_2) fixation*	X		X			
4. *Chemical composition*				X		
Macronutrient content				X		
Enzyme content						
Acetyl choline esterase					X	X
Phosphatase	X			X		
Nitrate reductase	X					
Toxic organics and metals content	X	X	X	X	X	X
5. *Flesh tainting*					X	X
6. *Histopathology*					X	X
7. *Condition factor*						X

Table 4-2
Use of Captive Organisms in Biomonitoring and Toxicity Tests

	Organism					
Type of Test	Phyto-plankton	Zoo-plankton	Peri-phyton	Macro-phyton	Macro-invert	Fish
In Situ Tests						
1. *Bioaccumulation*						
Toxic metals	X	X	X	X	X	X
Pesticides (organics)	X	X	X	X	X	X
Flesh tainting					X	X
2. *Toxicity tests*						
Acute toxicity					X	X
Histopathology					X	X
Histochemistry			X	X	X	X
Choline esterase						X
In-plant Tests (Effluents)						
1. *Bioaccumulation*						
Toxic metals			X	X	X	X
Pesticides (organics)			X	X	X	X
Flesh tainting					X	X
2. *Toxicity tests*						
Acute toxicity	X	X			X	X
Low-level responses						X
(behavioral responses)					X	X
Histopathology					X	X
3. *Biostimulatory tests*						
Algal growth response (AGP)	X		X			
Laboratory Tests						
1. *Bioaccumulation*						
Toxic metals	X	X	X	X	X	X
Pesticides (organics)	X	X	X	X	X	X
Flesh tainting					X	X
2. *Toxicity tests*						
Acute toxicity	X	X			X	X
Chronic toxicity	X	X	X	X	X	X
Histopathology					X	X
Low-level responses						X
(behavioral responses)					X	X
3. *Biostimulatory tests*						
Algal growth response (AGP)	X		X			

   B. Mutagenicity
       (1) Carcinogenicity
       (2) Teratogenicity
   C. Biostimulatory properties
       (1) Inorganic nutrients (N,P, and so on)
       (2) Degradable organic compounds

2. Properties of Communities of Indigenous Aquatic Organisms
   A. Standing crop—The numbers of and biomass (size, weight) of organisms present.
   B. Community structure—The kinds (and numbers) of species of organisms present (taxonomic composition), including the relative abundance of each kind and overall community structure and stability, and species diversity
   C. Community metabolism and condition—The rate of physiological processes such as photosynthesis (productivity), nitrogen fixation, and respiration; the bioaccumulation of toxic substances; and the occurrence of disease and histopathological conditions, parasitism, and flesh tainting.

This report deals only with programs that determine toxicity and other effects of pollutants on aquatic organisms.

Historical Review of Federal Biomonitoring Programs (Pre-EPA)

The legislative mandates and/or authority for biological monitoring contained in Public Law 95-217 are based on a concept entirely different from the approach taken under section 4(c) of the FWPA of 1956 (Public Law 660) and requires the collection of biological data because of its intrinsic value in determining the biological integrity of surface waters. However, the collection and processing of basic data on aquatic life under Public Law 660 were much more highly coordinated nationally than now under Public Law 95-217.

The biomonitoring program under Public Law 660 consisted of three types of activities: (1) National Water Quality Network (fixed-station, long-term ambient water quality monitoring), (2) Comprehensive River Basin Projects (long-term, intensive basinwide surveys), and (3) Enforcement studies (national laboratory, short-term intensive surveys).

National Water Quality Network (National Water Pollution Surveillance System)

Basic data on the species composition and abundance of aquatic life in the surface waters of the United States were first collected on a nationwide scale within the Federal Water Pollution Control Program by the National Water Quality Network (NWQN), established in 1956 under section 4(c) of Public Law 660 to determine the quality of surface waters used for domestic water supplies. By 1963 this network had grown to approximately 150 stations located throughout the forty-eight contiguous states and Alaska (figure 4-2a). The biological program was limited to phytoplankton counts and identifications during the first two years of operation, but gradually it expanded until 1963, when it included the collection of data on zooplankton, periphyton, macroinvertebrates, and fish (table 4-3).

The responsibility for the operation of the National Water Quality Network (later named the Water Pollution Surveillance System) was transferred to the regional offices in 1968, and the operation of the system was decentralized. Administration of the laboratory in which the samples were analyzed, now known as the Environmental Monitoring and Support Laboratory (EMSL-Cincinnati), was transferred to the EPA Office of Research and Development.

Comprehensive River Basin Projects

Laboratories located in each of the major river basins in the United States (figure 4-2b) carried out interdisciplinary, long-term, basinwide intensive studies to determine cause-and-effect relationships involving point and nonpoint sources of pollution in the basin. Data were collected on the physical, chemical, and biological integrity of water.

National Field Investigations Center (Cincinnati)

The National Field Investigations Center at Cincinnati was a field arm of the Enforcement Office of the Federal Water Pollution Control Administration Program. This center carried out short-term, intensive chemical and biological studies at a relatively small number of stations designed to provide data suitable for enforcement actions against point sources of pollution. These studies were heavily oriented toward biological effects.

Current Biomonitoring Programs

The EPA water monitoring programs are operated by the regional and research field laboratories, states, and other agencies, under the guidance of the EPA headquarters program offices (figure 4-3). Technical support for the field programs is provided by the various EPA research laboratories. The biomonitoring methods development, standardization, and quality assurance program is carried out at the Environmental Monitoring and Support Laboratory at Cincinnati (figure 4-4). The EPA now recognizes three basic types of monitoring: ambient (long-term) comonitoring. [National Pollution Discharge Elimination System (NPDES) Permit] compliance monitoring, and intensive (short-term)

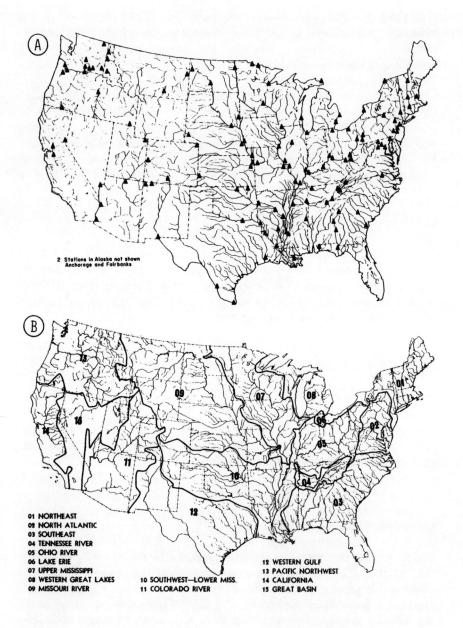

Figure 4-2. (*a*) Sampling Stations, National Water Quality Network (Water Pollution Surveillance System); (*b*) Major River Basins of the United States

Table 4-3
Biomonitoring Data Collected by the National Water Quality Network

Community	Sampling Period	Number of Stations	Sampling Method[a]	Type of Data
Phytoplankton	1956-1968	150	3-liter grab	S-R counts (units/ml) and species ID Diatom species, percent composition
Zooplankton	1956-1968	150	3-liter grab	Counts (org/l) and ID to genus
Periphyton	1964-1968	80	Glass slides in floating sampler	Cell density (cells/mm^2) and ID to species Diatom species, percent composition
Macroinvertebrate	1963-1968	40	Multiplates, rock-filled baskets, bottom grab	Counts and ID, organisms/sample or organisms/m^2
Fish	1963-1965	20	Electric shocker	Species composition and biomass

[a]For methods see C.I. Weber, *Methods of Collection and Analysis of Plankton and Periphyton Samples in the Water Pollution Surveillance System,* A&D Report No. 19 (Cincinnati: Department of Health, Education, and Welfare, 1966), and *Biological Field and Laboratory Methods for Measuring the Quality of Surface Waters and Effluents* (Cincinnati: EPA, 1973), EPA-670/4-73-001.

surveys. The ambient biomonitory program is conducted primarily by the states and the U.S. Geological Survey, whereas the compliance biomonitoring and intensive surveys are conducted by the EPA regional programs and by the states which have NPDES permitting authority.

Intramural EPA Biomonitoring Program

National Water Biomonitoring Program. The Monitoring and Data Support Division (MDSD), EPA Office of Water and Hazardous Materials, has developed two different national biomonitoring strategies, the Model State Water Monitoring Program (EPA 1975) and the Basic Water Monitoring Program (EPA 1977). The biological parameters described in these documents are listed in tables 4-4 and 4-5.

The approaches discussed in these reports are for the optional use by state programs in the MDSD-designed and supported National Water Pollution Surveillance System, operated for the EPA by the states, using section 106(e) funds, and the U.S. Geological Survey (USGS). This network contains approximately 165 stations, 110 of which are operated by USGS, and the remainder are operated by the states.

Regional Programs. The current biomonitoring programs in the EPA regional offices lack coordination at the national level. The size and expertise of the

38 Biological Monitoring for Environmental Effects

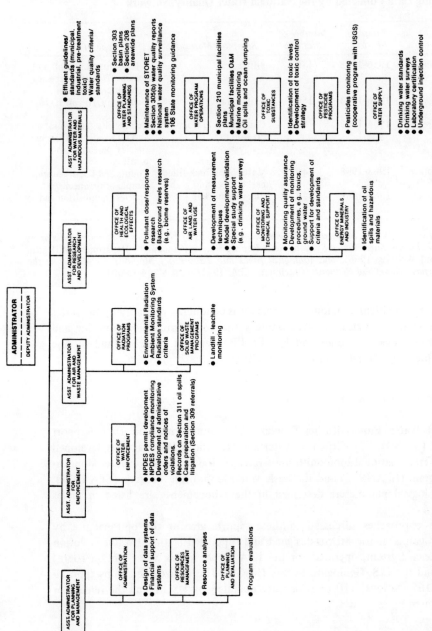

Source: *Basic Water Monitoring Program* (Washington: Environmental Protection Agency, 1977), EPA 440/9-76-025.

Figure 4-3. Environmental Protection Agency, Water Monitoring Activities

Federal and State Biomonitoring Programs

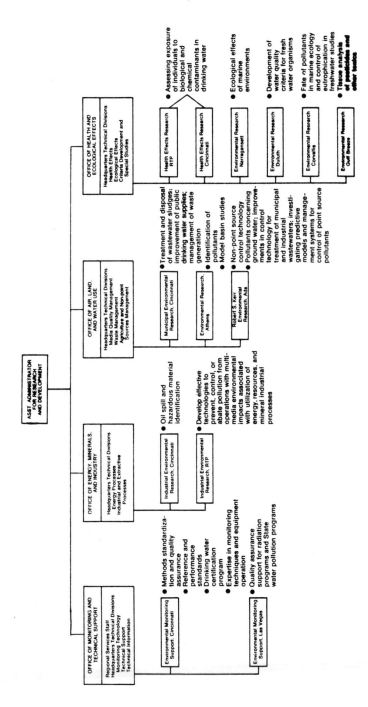

Source: *Basic Water Monitoring Program* (Washington: Environmental Protection Agency, 1977), EPA 440/9-76-025.

Figure 4-4. Water Monitoring Program Support Provided by the EPA Office of Research and Development

Table 4-4
Biological Parameter List, Sampling Frequencies, and Priority for Proposed Biomonitoring Programs

(a) Model State Water Monitoring Program

Community	Parameter	Priority[a]	Collection and Analysis Mathematics[b]	Sampling Frequency[c]
Plankton	Counts and identification	1	Grab samples	Once each—in spring, summer, and fall
	Chlorophyll *a*	2		
	Biomass as ashfree weight			
Periphyton	Counts and identification	1	Artificial substrates	Minimally once annually during periods of peak periphyton population
	Chlorophyll *a*	2		
	Biomass as ashfree weight	2		
Macrophyton	Areal coverage	2	As circumstances prescribe	Minimally once annually during period of peak macrophyton population density and/or diversity
	Identification	2		
	Biomass as ashfree weight	2		
Macroinvertebrate	Counts and identification	1	Artificial and natural substrates	Once annually during periods of peak macroinvertebrate population density and/or diversity
	Biomass as ashfree weight	2		
	Flesh tainting	2		
	Toxic substances in tissue[d]	2		
Fish	Toxic substances in tissue[d]	1	Electrofishing or netting	Once annually during spawning runs or other times of peak fish population density and/or diversity
	Counts and identification	2		
	Biomass as wet weight	2		
	Condition factor	2		
	Flesh tainting	2		
	Age and growth	2		

(b) Basic Water Monitoring Program

	Community of Aquatic Organisms			
	Plankton	Periphyton	Macroinvertebrates	Fish/Shellfish
Parameters				
Counts			X	
Species identification			X	X
Biomass (ashfree weight)		X		
Chlorophyll *a*	X	X		
Toxic substances				X[e]
Habitat Types				
Rivers		X	X	X
Lakes	X	X	X	X
Estuaries	X			X

Table 4-4 continued

	Community of Aquatic Organisms			
	Plankton	Periphyton	Macroinvertebrates	Fish/Shellfish
Sampling Methods Sampling season Sampling frequency Sampling method	6/15-9/15 Monthly grab	6/15-9/15 Once annually Glass slides Floating Sampler	6/15-9/15 Once annually Hester-Dandy Multiplace	6/15-9/15 Once annually
Number of replicate samples		3	3	3

[a]Priority: 1—minimum program; 2—Add as soon as capability can be developed.

[b]See EPA biological methods mannual.

[c]Keyed to dynamics of community.

[d]See *Analysis of Pesticide Residues in Human and Environmental Samples*, CSEPA, Perrine Primate Research Laboratories, Perrine, Fla., 1970, and *Pesticide Analytical Manual* (Washington: Department of Health, Education, and Welfare).

[e]Fish tissue analysis is a specific requirement of the basic ambient monitoring program.

Table 4-5
Biological Parameter Units

Community	Parameter	Units
Plankton	Counts	Numbers/ml by genus and/or species
	Chlorophyll *a*	mg/m^3
	Biomass (ashfree dry weight)	mg/m^3
Periphyton	Counts	Numbers/mm^2
	Chlorophyll *a*	mg/m^2
	Biomass (ashfree weight)	mg/m^2
	Autotrophic index	$\dfrac{\text{Ashfree weight } (mg/m^2)}{\text{Chlorophyll } a \ (mg/m^2)}$
Macrophyton	Areal coverage	Maps by species and species associations
	Biomass (ashfree weight)	g/m^2
Macroinvertebrate	Counts	Grab—number/m^2 Substrate—number/sampler
	Biomass	g/m^2
	Toxic substances	mg/kg
Fish	Toxic substances	mg/kg
	Counts	Number/unit of effort, expressed as per shocker hour or per 100 ft of a 24-hour net set
	Biomass (wet weight)	Same as counts
	Condition	$K(TL) = \dfrac{10^5 \times \text{weight in grams}}{L^3 \text{ (length in mm)}}$

Source: *Model State Water Monitoring Program* (Washington: Environmental Protection Agency 1975).

biology staff and the scope of the biology programs vary greatly from region to region (table 4-6), depending on the interpretation of the regional administrator and the surveillance and analysis division director, regarding the relative importance of biological data in responding to regional needs and the program guidance provided by EPA headquarters. The regions conduct intensive surveys, industrial and municipal effluent bioassays, sediment oxygen demand measurements, and determine the biological effects of ocean-disposed wastes and heated effluents. They also review Environmental Impact Statements and 316(b) applications. In all but region IV, the biology staff is generally too small to carry out an effective biomonitoring program (table 4-6).

In 1976 the regional programs operated 650 stations, collected approximately 3,600 samples, and had back data on 9,572 stations and 117,157 samples (tables 4-7 and 4-8) (Weber 1976b).

EPA National Enforcement Investigations Center (NEIC) (Denver). A field arm of the Office of Enforcement, the Center has an active biological program and conducts intensive field surveys, effluent bioassays, and other biological studies for regions lacking strong internal biological monitoring programs.

Research Programs. The Large Lakes Research Program is the largest single, current R&D biomonitoring program. This program maintains 200 stations and collects approximately 6,400 samples per year. Data are collected on phytoplankton, zooplankton, and toxic substances in fish tissues. Biological field and laboratory data are also generated by Environmental Monitoring and Support Laboratory (EMSL)-Cincinnati, EMSL-Las Vegas, and the environmental research laboratories at Duluth, Minnesota, Narragansett, Rhode Island, Gulf Breeze, Florida, and Corvallis, Oregon.

Table 4-6
Environmental Protection Agency Regional
Biomonitoring Programs

Region	Approximate Numbers of Biologists
I	2
II	3
III	3
IV	13
V	3
VI	2
VII	1
VIII	5
IX	1
X	2

Table 4-7
State Programs: Current Rate of Biological Data Acquisition

		Number of Samples per Year				
State[a]	Number of Stations	Plankton	Periphyton	Macroinvertebrates	Fish	Total
Alabama	25			50	18	68
Arkansas	25	133	41	37		211
California	80	370	60	60	150	640
Connecticut	30	200	100	80	30	410
Delaware	10			10		10
District of Columbia	3	36	36	36	36	144
Florida	239	75	63	516	0	654
Idaho	40			480		480
Indiana	200	700	50	15	15	780
Kentucky	50	300	200	250	100	850
Maryland	200			200		200
Michigan	250	611	125	583	450	1,769
Minnesota	40	60	4	30		94
Montana	100		200	200		400
Nevada	4	8	8			16
New Jersey	210	400		230		630
New Mexico	20	50		40		90
New York	80			100		100
North Carolina	128	156	156	208	104	624
Ohio	60			400		400
Pennsylvania	1,800		200	2,400	200	2,800
South Carolina	90	360	90	90	180	720
Tennessee	45	540	540	540	1,080	2,700
Texas	130	145		130		275
Vermont	100	360	24	360	24	768
Virginia	340	1,000		240		1,240
West Virginia	20			27		27
Wisconsin	200	175	70	685	20	950
Total	4,519	5,679	1,967	7,997	2,407	18,050

Source: C.I. Weber, *Feasibility Study for a Centralized Biological Data Management System (BIO-STORET)*. (Cincinnati, Ohio: Environmental Protection Agency 1976).

[a]The absence of any state from this list does not necessarily mean that it lacks a biological monitoring program.

Extramural Environmental Protection Agency Programs

EPA Grantees and Contractors. Agency grantees and contractors collect large amounts of field and laboratory data on aquatic organisms that would be of value to EPA headquarters and regional programs and states if these data were available in a national, computerized, biological data management system.

State Biomonitoring Programs. With the aid of EPA section 106(e) grants, state water pollution control agencies in 1976 maintained approximately 4,500

Table 4-8
State Programs: Back Data from Biomonitoring Programs[a]

| State[a] | Number of Stations | Number of Samples |||||
		Plankton	Periphyton	Macroinvertebrates	Fish	Total
Alabama	25	–	–	45	12	57
Arkansas	130	2,100	–	2,100		4,200
California	400	1,850	–	300	750	2,900
Connecticut	30	250	250	250	60	810
Delaware	30			60		60
Florida	150	40	20	950	–	1,010
Idaho	40	–	–	300	–	300
Maryland	250			200	50	250
Michigan	1,500	1,700	400	4,350	550	7,000
Minnesota	75	400	20	200	50	670
New Jersey	250	850	–	200	–	1,050
New Mexico	50	250	–	250	–	500
New York	425	–	–	875	–	875
North Carolina	128	150	150	200	28	528
Ohio	153	–	–	350	–	350
Pennsylvania	4,650	–	400	7,500	400	8,300
South Carolina	57	515	150	150	150	965
Tennessee	45	540	540	540	1,080	2,700
Texas	230	684	–	961	–	1,645
Vermont	400	850	30	2,000	50	2,930
Virginia	108	–	–	432	–	432
West Virginia	20	36	–	27	–	63
Wisconsin	1,600	400	150	1,400	50	2,000
Total	10,746	10,615	2,110	23,640	3,230	39,595

Source: C.I. Weber, *Feasibility Study for a Centralized Biological Data Management System (BIO-STORET)* (Cincinnati, Ohio: Environmental Protection Agency, 1976).

[a]The absence of any state from this list does not necessarily mean that it lacks a biological monitoring program.

stations and collected about 18,000 biological samples per year. Back data existed for approximately 40,000 biological samples from 11,000 stations. (See tables 4-7 and 4-8). The 1978 state programs are described in table 4-9. The evaluation of the biological data, together with the physical and chemical water quality data, is used by some states to prepare the biennial reports to EPA required under section 305(b), Public Law 95-217. These reports describe the extent to which the navigable waters of the states provide for the protection and propagation of aquatic life. However, current guidelines provided to the states by the EPA for the preparation of the section 305(b) reports do not require mention of biological conditions. As a result, most reports do not include comments on the biological integrity of the state's waters.

Other Organizations

In addition to the work of the EPA, many other federal, interstate, and private agencies are engaged in studies of the effects of pollution on aquatic life. Much of this work is done with pass-through EPA funds and constitutes an extension of its program. It would be to the advantage of the EPA to work more closely with other agencies and to capture their data in a central data processing facility to permit pooling and evaluation of related data, where possible, provide timely access to all data for program planning and evaluation, and reduce or eliminate duplication of effort.

Federal Programs. Other federal programs that have biomonitoring programs include:

1. Tennessee Valley Authority
2. U.S. Department of the Interior
    U.S. Geological Survey
    Bureau of Reclamation
    Bureau of Land Management
    Office of Biological Services, U.S. Fish and Wildlife Service
3. National Oceanic and Atmospheric Administration (NOAA)
4. U.S. Army Corps of Engineers
5. Nuclear Regulatory Commission
6. Department of Energy

Other programs include state water pollution control agencies, interstate river basin commissions, and area and basin planning commissions (section 208). Available information on the number of stations operated and samples collected by other federal programs is listed in tables 4-10 and 4-11.

Biomonitoring Methods Development, Standardization, and Quality Assurance

Within the EPA, the Aquatic Biology Section, the Environmental Monitoring and Support Laboratory (EMSL) at Cincinnati, has the responsibility for developing, evaluating, and standardizing biomonitoring methods and quality assurance procedures, in support of the regional and state biomonitoring programs.

Program areas include project planning, methods and sample collection and preparation, identification and enumeration of organisms, measurements of biomass and metabolic growth rates, measurements of effluent toxicity, and the bioaccumulation of toxic substances in tissues. The Aquatic Biology Section

Table 4-9
State Biomonitoring Programs, 1978

| Region | State | Staff | Number of Stations ||||| Number per Year ||
			Plankton	Periphyton	Macro-invertebrates	Section 314	Toxics in Fish	Effluent Bioassay (Static/FT)	Intensive Surveys
I	ME				11	250			
	MA				9				x*
	VT					x			x
II	NJ		Marine	x	x	x			
	NY	3			224		224		x
III	DE	1	15		15		20	2-3/0	
	MD	3	75		244		x	10/0	
	PA	10			285			12/3-5	
	VA	11			120		40	6/0	
	WV	1½	4	4	22	28			
IV	AL	6			26		26	x/0	
	FL	30	76	76	247	247	247	x	
	GA	4	46	46	46				
	KY	4	7	7	7		7		
	MS	2	16	16	16		16	Plan FT	
	NC	4	256	256	256	x	256	Plan AGP	
	SC	6	45	45	45		45	Plan AGP	
	TN	11	21	21	21		21		
V	IL	8			50				10
	IN	5	Plan	Plan	Plan	x		x/x	x
	MI	5	41	41	41			x/x	
	MN	5		20	20				
	OH	5		40	40		Plan		
	WI	4		29					

	State									
VI	AR	4							23		1
	LA	2	Plan	40	Plan	40	Plan	40	44		
	NM	3	Plan	44	Plan	44	Plan	44	7		
	OK	2		9		9		9	22		
	TX	15		22		75		22	75	x	13	21
VII	IA	3		36				36			
	KS	3		40				40			
	MO	1						50			
VIII	CO	4	Plan	126	Plan	126	Plan	126			
	MT	2									
	ND	9							x		
	UT	1									
IX	(No information)
X	AK		(Marine)
	ID				x		x				
	WA				x		x				

"x" indicates some activity, but level is unknown.
Ft = Flow-through
AGP = Algal growth potential test

Table 4-10
Federal Programs: Current Rate of Biological Data Acquisition

	Number of Stations	Number of Samples per Year				
		Plankton	Periphyton	Macro-invertebrates	Fish	Total
EPA						
Region: I						
II	32	24	–	100	24	148
III	238	440	–	150	90	680
IV	100	400	150	250	50	850
V	100	780	–	520	–	1,300
VI	200	303	91	191	25	610
VII	6	–	–	24	–	24
VIII	15	?	?	?	?	
IX						
X	60	–	360	360	–	720
R&D Corvallis Lab	4	96	–	396	–	492
R&D Grosse Ile Lab	200	4,800	–	400	1,200	6,400
R&D Cincinnati Lab (EMSL)	10	100	–	100		200
Department of Agriculture National Forest Service						
Prineville, OR	40	?	?	?	?	?
Albuquerque, NM	50	40	20	220	–	280
Alamogordo, NM	9	–	–	18	–	18
Durango, CO	12	–	–	–	20	20
Russelville, AK	20	–	–	–	20	20
Provo, UT	40	–	–	500	–	500
Total	1,136	6,983	621	3,229	1,429	12,262

Source: C.I. Weber, *Feasibility Study of a Central Biological Data Processing System* (Cincinnati, Ohio: Environmental Protection Agency, 1976).

is also responsible for developing methods for the storage, retrieval, analysis, and interpretation of biological data and for biological quality assurance.

All communities of aquatic organisms are considered, including the phytoplankton, zooplankton, periphyton, macrophyton, macroinvertebrates, and fish. A biological methods manual was published in 1973, and the second edition is nearing completion (1979). Publications include protocols for the evaluation of biological quality assurance programs, manuals for the identification of aquatic organisms, results of the evaluation and comparison of sampling methods, compilations of data on the pollution tolerance of common aquatic organisms, and methods for measuring effluent toxicity. The specifications for a computerized biological data management system and a coded master list of 11,000 aquatic species also have been published.

Table 4-11
Federal Programs: Back Data from Biomonitoring Programs

	Number of Stations	Number of Samples				
		Plankton	Periphyton	Macroinvertebrates	Fish	Total
EPA						
NEIC-Denver	350	150	100	300	200	750
Region I						
II	515	1,020	–	1,500	30	2,550
III	20	520		740	30	1,290
IV	250	2,104	906	1,611	240	4,861
V	4,000	3,000	300	3,700	–	7,000
VI						
VII						
VIII	80	15	15	180	1,000	1,210
IX						
X						
Grosse Ile Lab	345	8,000	400	400	6,200	15,000
Las Vegas Lab	2,400	1,850	–	500	–	2,350
Cincinnati (EMSL)	150	75,000	500	1,000	100	76,600
U.S. Army Corps of Engineers						
Buffalo, N.Y.	100	100	100	100	100	400
U.S. Department of Agriculture National Forest Service						
Prineville, OR	40					600
Alamogordo, NM	200	240	40	3,200	–	3,480
Albuquerque, NM	17	–	–	306	–	306
Durango, CO	45	–	–	110	–	110
Russelville, AK	20	–	–	–	–	600
Provo, UT	40	–	–	500	–	500
Total	8,572	91,999	2,361	14,147	7,900	117,607

Source: C.I. Weber, *Feasibility Study for a Centralized Biological Data Management System* (Cincinnati, Ohio: Environmental Protection Agency, 1976).

With the cooperation of the Quality Assurance Branch, EMSL, several biological reference materials have also been prepared. Those now available include:

1. Chlorophyll extract in 90 percent aqueous for the spectrophotometric method (performance evaluation and quality control). The sample contains chlorophylls *a, b,* and *c* and pheophytin *a.*
2. Chlorophyll extracts in 90 percent aqueous acetone for fluorometric methods (calibration, performance evaluation, and quality control). Three solutions are available: Two solutions contain purified chlorophyll *a* at different concentrations, and one solution contains a mixture of chlorophyll *a* and pheophytin *a.*

3. Simulated phytoplankton sample for microscope calibration and plankton counting. The sample consists of an aqueous suspension of latex spheres (15,000/ml), varying in size from 4 to 20 μm in diameter.

Additional materials in preparation include the following:

1. Reference toxicants for bioassays: sodium dodecyl sulfate, diquat, and sodium pentachlorophenate
2. ATP samples: four concentrations ranging from 10^{-6} μg/ml to 1 μg/ml
3. Plankton sample for S-R counts and identification
4. Periphyton sample for S-R counts and identification
5. Diatom hyrax mounts, for species proportional counts
6. Diatom reference specimens (hyrax mounts of identified specimens)
7. Macroinvertebrate grab samples for sorting, counting, and identification
8. Macroinvertebrate reference specimens (collections of identified specimens)

Summary and Recommendations

This chapter outlines the federal biological monitoring programs for implementing the objectives of the 1972 and 1977 amendments to the FWPCA.

Legislation provides authorization for the EPA, states, and other federal agencies to conduct comprehensive biological monitoring programs to determine the effects of pollution on recreational water and aquatic life. Current EPA guidelines for preparation of section 305(b) reports by states do not, however, require mention of biological conditions of the state's waters. It is recommended that this oversight be corrected by providing useful instructions and guidelines to states for obtaining information and preparing an annual assessment of the biological integrity of states' water resources.

The current level of staffing of biologists in EPA Regional Offices, Surveillance and Analysis Divisions with biologists is generally inadequate to conduct an optimal biological monitoring program. It is recommended that the staffing levels of the EPA and state biomonitoring programs be raised at least to the "minimum" level recommended in the EPA report entitled *Model State Waters Monitoring Program*.

References

Beck, W.M. 1977. *Environmental Requirements and Pollution Tolerance of* Chironomidae. Environmental Monitoring and Support Laboratory. Cincinnati, Ohio: Environmental Protection Agency, EPA 600/4-77-024.

Harris, T. 1978. *Environmental Requirements and Pollution Tolerance of* Trichoptera. Environmental Monitoring and Support Laboratory. Cincinnati, Ohio: Environmental Protection Agency.

Hubbard, M.D., and Peters, W.L. 1978. *Environmental Requirements and Pollution Tolerance of* Ephemeroptera. Environmental Monitoring and Support Laboratory. Cincinnati, Ohio: Environmental Protection Agency, EPA 600/4-78-061.

Klemm, D.J. 1979. *Freshwater Leeches* (Annelida: Hirudinea) *of North America.* Environmental Monitoring and Support Laboratory. Cincinnati, Ohio: Environmental Protection Agency.

Lewis, P.A. 1974. *Taxonomy and Ecology of* Stenonema *Mayflies* (Heptageniidae; Ephemeroptera). Methods Development and Quality Assurance Laboratory. Cincinnati, Ohio: Environmental Protection Agency.

Lowe, R.L. 1974. *Environmental Requirements and Pollution Tolerance of Freshwater Diatoms.* Environmental Monitoring and Support Laboratory. Cincinnati, Ohio: Environmental Protection Agency, EPA 670/4-74-005.

Mason, W.T., Jr. 1968. *An Introduction to the Identification of Chironomid Larvae.* Cincinnati, Ohio: Division of Pollution Surveillance, Federal Water Pollution Control Administration, U.S. Department of Interior.

Nacht, L., and Weber, C.I. 1976. *BIO-STORET Final Design Specification.* Environmental Monitoring and Support Laboratory. Cincinnati, Ohio: Environmental Protection Agency.

Peltier, W. 1978. *Methods for Measuring the Acute Toxicity of Effluents to Aquatic Organisms,* 2d ed. Environmental Monitoring and Support Laboratory. Cincinnati, Ohio: Environmental Protection Agency, EPA 600/4-78-012.

Schuster, G.A., and Etnier, D.A. 1978. *A Manual for the Identification of the Caddisfly Genera* Hydropsyche *Pictet and* Symphitopsyche *Ulmer in Eastern and Central North America (Trichoptera: Hydropsychidae).* Environmental Monitoring and Support Laboratory. Cincinnati, Ohio: Environmental Protection Agency.

Surdick, R.F., and Gaufin, A.R. 1978. *Environmental Requirements and Pollution Tolerance of* Plecoptera. Environmental Monitoring and Support Laboratory. Cincinnati, Ohio: Environmental Protection Agency, EPA 600/4-78-062.

Environmental Protection Agency. 1974. *Water Quality and Pollutant Source Monitoring.* Monitoring and Data Support Division, Office of Water and Hazardous Materials, Washington. Fed. Reg. 39(168):31500-31505.

_____. 1975. *Model State Water Monitoring Program.* Monitoring and Data Support Division, Office of Water and Hazardous Materials, Washington.

_____. 1977. *Basic Water Monitoring Program.* Monitoring and Data Support Division, Office of Water and Hazardous Materials, Washington, EPA 440/9-76-025.

Weber, C.I. 1966a. *A Guide to the Common Diatoms at Water Pollution Surveillance System Stations.* Cincinnati, Ohio: Division of Pollution Surveillance, Federal Water Pollution Control Administration, U.S. Department of Health, Education and Welfare.

_____. 1966b. *Methods of Collection and Analysis of Plankton and Periphyton Samples in the Water Pollution Surveillance System*. Cincinnati, Ohio: A&D Report No. 19, Division of Pollution Surveillance, Department of Health, Education and Welfare.

_____. 1973a. *Biological Field and Laboratory Methods for Measuring the Quality of Surface Waters and Effluents*. Methods Development and Quality Assurance Laboratory. Cincinnati, Ohio: Environmental Protection Agency, EPA-670/4-73-001.

_____. 1973b. "Biological Monitoring of the Aquatic Environment by the U.S. Environmental Protection Agency." In *Biological Methods for the Assessment of Water Quality*, eds., J. Cairns, Jr., and K.L. Dickson, ASTM Special Publication No. 528. Philadelphia: American Society for Testing and Materials, pp. 46-60.

_____. 1976a. *BIO-STORET Master Species List*. Environmental Monitoring and Support Laboratory. Washington: Environmental Protection Agency. (Second edition in press.)

_____. 1976b. *Feasibility Study for a Central Biological Data Management System (BIO-STORET)*. Environmental Monitoring and Support Laboratory. Cincinnati, Ohio: Environmental Protection Agency.

5
Biological Monitoring to Provide an Early Warning of Environmental Contaminants

Kenneth L. Dickson, David Gruber, Christine King, and *Kenneth Lubenski*

Biological monitoring approaches have at least two uses in protecting aquatic ecosystems from damage from potentially hazardous chemical substances. First, the responses of aquatic organisms upon exposure to a chemical substance can be used to estimate the hazards associated with the use of the chemical substance. Second, biological monitoring systems can be used to continuously monitor the quality of the aquatic ecosystem and detect the presence of harmful environmental contaminants. By using information on the environmental fate of a chemical substance along with data on its effects, an assessment of hazard can be established. The environment can then be protected by means of decisions to ban or limit the use of potentially hazardous materials.

A description of the biological monitoring techniques under investigation by the authors is presented. Living organisms serve as indicators of environmental quality by monitoring the ventilatory behavior, activity, and locomotor behavior of fish.

Introduction

Biomonitoring has some remarkably interesting relationships with medieval kings, coal miners, and Colonel Sanders, of Kentucky Fried Chicken fame. The roots of the concept of biomonitoring might well originate with medieval kings who guarded against poisoning by having their food and wines tasted by guards prior to consumption. Likewise, the concepts of biomonitoring were nurtured by the practice of miners who took caged canaries into their mines to monitor the presence of deadly gases. Finally, while it may seem ludicrous on first inspection to associate biomonitoring with Colonel Sanders, it is not unreasonable to relate the recent explosion of interest in biomonitoring to the phenomenal success of Kentucky Fried Chicken, which did not become franchised until the founder was 66 years old. While biomonitoring is not currently experiencing the same kind of growth, it is evident that it, too, has come of age.

The authors wish to express their gratitude to their respective granting agencies. These include the Manufacturing Chemists Association and the Office of Ecological Research, U.S. Army.

Roles of Biomonitoring

Other chapters in the book have related biological monitoring approaches in detecting environmental contaminants in air and interterrestrial systems. This chapter, however, examines the roles of biomonitoring from an aquatic viewpoint.

There are two roles that biomonitoring must play if we are to protect our aquatic resources from damage. (1) We must be able to provide a real-time (that is, continuous) assessment of present conditions. (2) We must use our biomonitoring technology to predict the effects of new chemical substances likely to reach aquatic ecosystems.

Conveniently, aquatic organisms serve as integrators of their total environment, and often they respond to extremely low concentrations of environmental contaminants. The challenge facing us is to develop the methods and technology to detect and use the responses of organisms to provide information about the presence and effects of these contaminants. In addition, we must develop the capability to do this in *real-time*. Here, real-time implies continuous monitoring and the ability to analyze responses in time to prevent ecological damage.

In addition, we must use our biomonitoring tools to predict the *future* ecological effects of chemical substances and utilize this knowledge to prevent these substances from reaching hazardous concentrations in the ecosystem. We must learn how to use laboratory studies on acute effects, chronic studies on growth and reproduction, bioconcentration experiments, and behavioral and physiological studies to evaluate chemical substances for their hazard as long-term environmental contaminants. The key questions are these: How do you assess the risk or hazard? What tests need to be conducted? Do all chemical substances have to be tested equally? Are there enough dollars and workforce to do the job? These are not easy questions to answer! However, it appears that biomonitoring has the potential to help answer many of them.

Work is underway on many fronts to develop real-time biological monitoring systems which provide an early warning of environmental contaminants. These have potential to partially fill the first role identified above. In addition, considerable effort is being made in the development of methods and testing protocols to assess the hazards associated with chemical substances *before* their introduction into the environment. These laboratory testing activities are designed to provide the information necessary to make informed decisions about the use of chemical substances in the environment.

Watershed Management Concept

Biological monitoring systems which provide an early warning of the presence of environmental contaminants are an integral part of a watershed management approach to protecting environmental quality (figure 5-1).

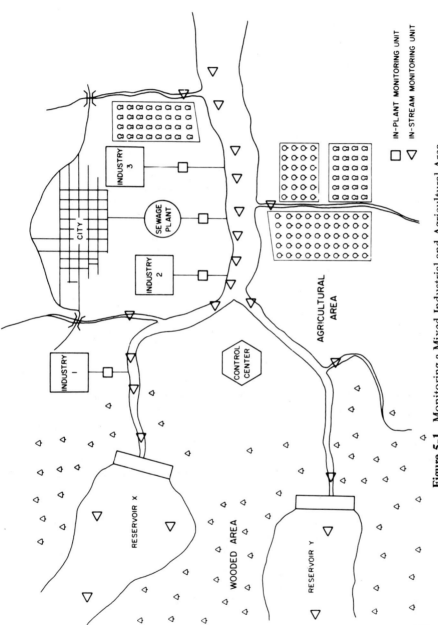

Figure 5-1. Monitoring a Mixed Industrial and Agricultural Area

Management of aquatic ecosystems requires a clear understanding of the goals to be achieved, appropriate information, and the means to achieve the goals. Control measures applied to aquatic ecosystems, in the absence of information on the condition of the system, are apt to be inappropriate and thus may overprotect the receiving system at some times and underprotect it at other times, since the ability of ecosystems to receive wastes is not constant. A major determinant of the effectiveness and efficiency of ecological quality control is the lag time in the feedback of information. If the lag is too great, the control measures may repeatedly overshoot or undershoot the desired goal.

Society's interest in pollution is centered on its effects on living organisms; therefore, it seems reasonable to develop and utilize biological pollution monitoring systems if we are to achieve our goal of environmental quality management. Most pollution monitoring programs have been oriented solely toward chemical and physical parameters. However, it is difficult, if not impossible, to predict the biological effects of a complex, continuously changing industrial waste from chemical-physical analyses alone. Damage may be caused by unsuspected contaminants: a janitor may dump a cleaning solvent down a drain, or a change in upstream conditions in combination with a certain effluent may produce toxic conditions. Since organisms act as integrators of the total environment, their response to complex sets of environmental conditions can be used as monitors of water quality. The development of biological monitoring systems has not kept pace with the development of chemical-physical monitoring systems. Continuous monitoring systems for analyzing dissolved oxygen, pH, conductivity, turbidity, total organic carbon, and specific ions via specific ion electrodes have been available for several years.

In essence, total monitoring systems (biological, chemical, and physical) are essential to obtain the information necessary to apply control measures to aquatic systems if we are to manage the systems to allow people to make full beneficial use of the receiving capacity of natural ecosystems. Neither the abilities of ecosystems to receive wastes nor the toxicity of effluents remains constant. Thus, if continuous biological, chemical, and physical information on the receiving system is not available, control measures applied to aquatic ecosystems are apt to be inappropriate, overprotecting the receiving system at times and underprotecting it at other times. The precision and efficiency of environmental quality control and the lag time in the information feedback loop are related. If the lag time is too great, the control measures may repeatedly overshoot and undershoot the desired goal, as a thermostat with too slow a response will first cause underheating, then overheating, of a house. We must go beyond rigid, arbitrary standards for disposal of wastes into the environment to a quality control system based on the day-to-day and week-to-week loading capacity of each system. To manage a complex ecosystem with many interlocking cause-effect pathways, we must determine the impact of all inputs into the system—industrial, municipal, agricultural, and so on. In addition, we must be able to monitor all these inputs, treating an entire drainage basin as an operating system and relating the effects so that a temporary increase in one component may require a temporary decrease in another component. The key to success in a large-scale systems approach to ecosystem management is rapid generation of biological, physical, and chemical information. Fairly rapid feedback techniques for chemical and physical information are already available. Biological information systems with comparable or nearly comparable lag times, however, need to be developed.

Basically, the quality control of an ecosystem will depend on the related, but not identical, biological monitoring systems. One of these will be the in-plant early-warning system that can be incorporated into the waste disposal system of an industry so that the presence of potentially lethal or harmful materials can be detected before the waste reaches the stream. The second would be an in-stream system which will give information regarding the state of health of various regions of an entire drainage basin. In short, optimal beneficial use of an ecosystem will depend on the rapidity with which ecological quality control assessments can be made of the impact of each waste on the receiving system itself and on representative species of the system used as biological sensors incorporated into the waste disposal of every plant.

Figure 5-1 illustrates how a system of such monitors might be organized in a watershed. The in-plant monitors would be placed at all major industries and sewage treatment facilities. The in-stream devices would be placed at strategic locations in the watershed itself. The data recorded at all these would be transmitted continuously or at regular intervals to a central location for analysis.

If warnings of a toxic discharge came from the monitoring units at Industry 3, it might undertake its own control measures; but in any event, additional control measures are available if a river management system is in operation. The control center might request Industry 1 and Industry 2 to hold their waste so as not to overload a system already strained by the toxic discharge from Industry 3. Alternatively, the control center could call for a release of water from reservoirs X and Y to dilute the waste from Industry 3.

Fish Biological Monitoring System for Point-Source Discharges

For the past eight years, a biological monitoring system has been under development in the aquatic ecology laboratory at Virginia Polytechnic Institute and State University. This system has been designed to continuously monitor industrial and municipal effluents and to provide an early warning of harmful substances in these discharges.

Public Law 92-500 requires biological monitoring of waste discharges. This may be accomplished by direct measurements within the receiving system itself or by using an aquatic species (for example, fish) in a container into which a mixture of waste and receiving water is introduced. Although at times this may be the only form of monitoring, more often chemical and physical monitors are associated with this system. However, it is difficult to assess the biological effects an effluent may have when chemical and physical monitors alone are used. In fact, single assessments are inadequate for precise loading calculations because there are fluctuations (which may be daily or even hourly) in the receiving system's ability to receive wastes. The effect a toxicant ultimately has on a community of organisms will also depend on the interactions that a toxicant has with other toxicants which already may be present in the receiving system. The toxicants may interact synergistically so that their combined toxicity is much higher than one would predict from their individual additive toxicities. On the other hand, some compounds may prove toxic at levels below the analytical capabilities of some instruments. Although biological monitors serve as useful integrators of effluent and receiving-system conditions, identification of the cause of deleterious biological effect will generally require accompanying chemical and/or physical information. Therefore, it is essential to have a monitoring program that includes all three components. In addition, the use of minicomputers in these systems will reduce the time required to generate useful information and greatly reduce the cost per unit of information as well.

The following discussion describes the biomonitoring system.

The biological sensor in the system is the bluegill sunfish (*Lepomis macrochirus*), an organism found throughout the United States. The fish range in size from 8 to 14 cm, weigh from 15 to 50 g, and either have been collected locally or obtained from several commercial suppliers. They have been maintained in dechlorinated municipal tapwater, fed a commercial food once a day, and generally kept on 12-hr light-dark photoperiods. Although it is feasible to monitor an assortment of behavioral parameters, previous research from this and other laboratories suggests that the breathing response is optimal for this application. Abnormal breathing responses rapidly appear in fish exposed to sublethal concentrations of toxicants and frequently well in advance of irreversible damage.

Electrical biopotentials are normally produced by the opercular and other muscular movements incorporated with fish "breathing." For 8- to 14-cm bluegills, these signals generally range from 20 to 50 μV. That these signals are produced and may be recorded has been previously established [1, 2, 3, 4, 5, 6, 7]. Signals are picked up through uninsulated, single-strand, stainless-steel wire electrodes sealed with epoxy resin to the inside of the tanks. In order to provide an input of low-voltage signals to a computer, these signals need to be amplified by from 1 to 5 × 10^5. This is accomplished by feeding the respiratory signals generated by each fish through relatively inexpensive noise-immune amplifiers developed in our laboratory. Detailed technical description of these amplifiers has previously been reported [4].

In this biological monitoring system, all instrumentation, including a small on-line computer, are housed within a 10 X 4 m trailer to facilitate on-site monitoring (figure 5-2). Figure 5-3 shows the layout of the inside of the trailer. The exposure tanks are housed within the trailer in a modularized facility. Each is constructed of clear Plexiglas, has a 5-liter capacity, and possesses sloping floors which facilitate tank cleaning (figure 5-4). Each tank is designed to contain and monitor one fish. The module employs twelve such tanks although it has been designed to capacitate up to forty-eight tanks. Figure 5-5 shows a diagram of the module housing the chambers in the trailer. Both the feeding and lighting schedules for the experimental tanks are controlled through computer logic.

Four chemical-physical parameters can also be monitored. Conductivity, dissolved oxygen, pH, and temperature can be monitored periodically by instrumentation interfaced to the computer. In this fashion, monitored values will be stored on a computer disk for future reference.

A PDP-11-VO3 computer is incorporated in the monitoring system. This computer system basically consists of an LSI-11 minicomputer, a dual floppy disk, and a Decwriter. This system enables us to rapidly and continuously monitor the individual signals of many fish while frequently assessing their breathing rates. A statistical package, programmed into the computer system, facilitates assessment of the data as frequently as desired and was developed in cooperation with the statistics department of Virginia Polytechnic Institute and State University [8]. Essentially, this procedure determines a range of acceptable breathing rates, called the *critical limits* for each fish, taking into consideration diurnal rhythmmicities. During the monitoring periods, data are assessed by comparing the sampled breathing rates to the set of critical limits previously determined for each fish. A value that is outside the critical limits will cause the computer system to immediately generate a warning signal. Furthermore, an alarm system could be initiated if a predetermined number of fish generate such warnings. The number of fish needed to generate an alarm system would be determined with respect to the nature of the effluent and its relative toxicity.

A graphic display depicting the results of a simulated chlorine spill is presented in figure 5-6. While the breathing rates of the fish not exposed to chlorine remained fairly constant, all four fish exposed to chlorine displayed a dramatic decrease in the frequency of breathing counts recorded. Had critical limits been determined for these fish, several low responses would have been indicated. This situation would have generated an alarm, and appropriate followup action would have been taken.

Biological monitoring systems can generate quick information concerning a waste toxicant level change. Since the monitored data are capable of being

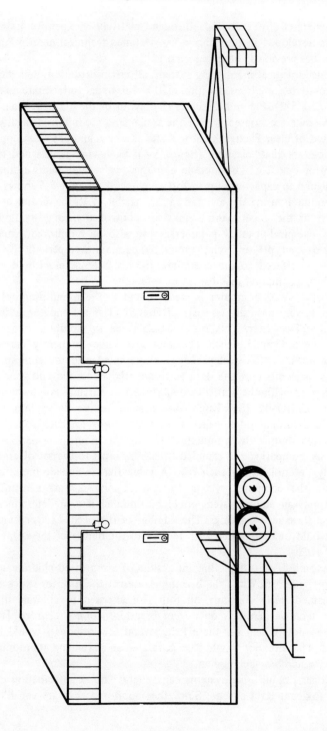

Figure 5-2. Mobile Biological Monitoring System

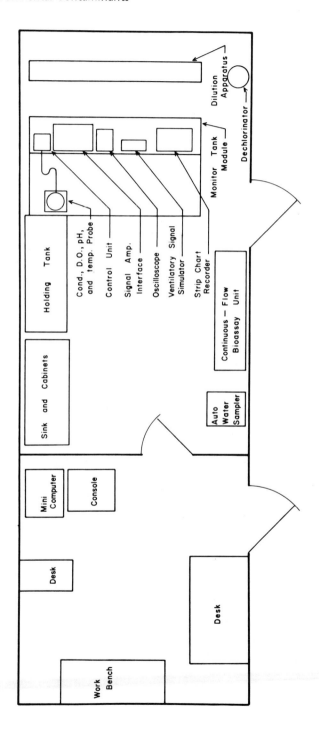

Figure 5-3. Internal Layout of the Biological Monitoring Trailer Facility at Virginia Polytechnic Institute and State University's Center for Environmental Studies

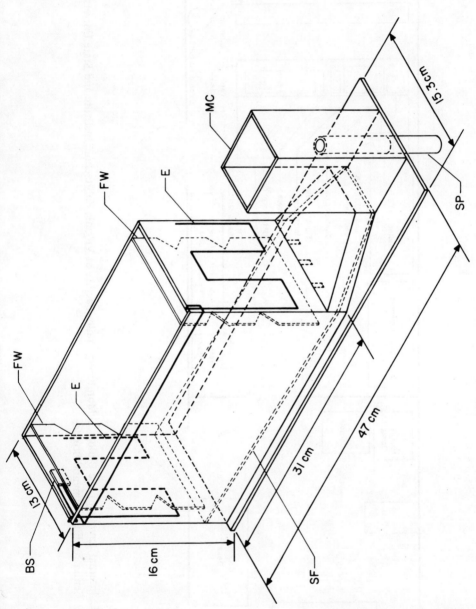

Figure 5-4. Plastic Exposure Tanks

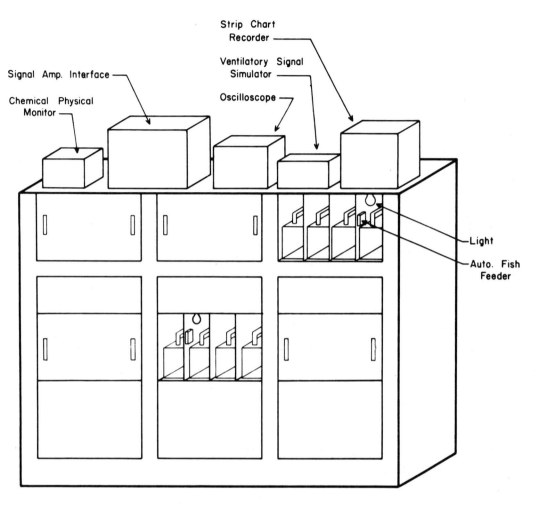

Figure 5-5. Module Housing Chambers: Fish Exposure and Monitoring

interfaced to computers, the lag time between detection and correction of problems may be dramatically reduced. Information generated from the biological monitoring system would also be capable of regulating waste treatment processes. This not only would curtail frequent costly and unnecessary overtreatments of effluent wastes but also would potentially prevent devastating undertreatments. Finally, if such systems were developed for a series of plants on a single river or lake, it would be possible to coordinate their discharges so as to prevent kills while maximizing the concentrations of their effluents periodically. Such an effort would certainly prove most beneficial to the ecosystem as well as to humans.

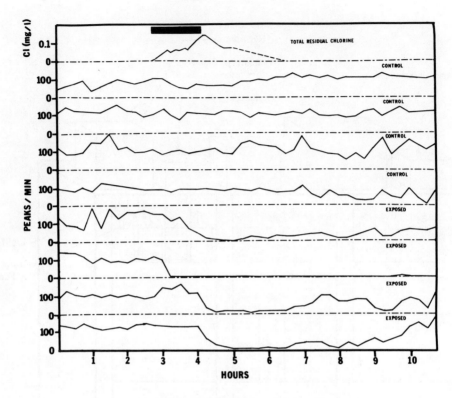

Figure 5-6. Fish Breathing Rates—Chlorine Exposure

Biological Monitoring Systems to Predict the Effects of Chemical Substances on Aquatic Life

In October 1976, the Toxic Substance Control Act (TSCA) became law. This law provides that no person may manufacture a new chemical substance or processs a chemical substance for a new use without obtaining clearance from the EPA. The Toxic Substance Control Act represents an attempt to establish a mechanism whereby the hazard to human health and the environment of a chemical substance can be assessed before its introduction into the environment. The challenge lies in how to assess the hazard in a realistic and economically feasible manner.

As might be expected, the enactment of TSCA has provided a powerful new stimulus to the development of testing procedures to evaluate the hazard to human health and to the environment associated with potentially toxic chemical substances. At present the EPA Office of Toxic Substances, many industries, and a number of researchers working in aquatic toxicology are

Environmental Contaminants

trying to develop reasonable approaches to accomplish the mandate of the law. If one considers that there are a minimum of 1,500 new chemical substances per year which must be evaluated under TSCA, the magnitude of the problem should be evident.

How is the hazard associated with a chemical substance evaluated? How does biomonitoring figure into this? Well, figure 5-7 shows that the judgment on hazard is based on four factors: (1) use; (2) physical, chemical, and biological properties of the chemical substance; (3) biological response; and (4) environmental concentration. The lower the ratio of the environmental concentration of the chemical to the lowest concentration causing an adverse biological effect, the less the hazard. The challenge for those working in the field of biomonitoring is to develop methods and technology enabling detection of an effect before damage occurs. It must be stressed that we need to develop methods and technologies that are cost-effective.

Fathead Minnow Biomonitoring System

It is evident that one of the most important bits of data needed to predict the hazard of a chemical substance to aquatic life is that concentration of the chemical above which there are observable biological effects and below which no effects are observed. This is the maximum acceptable toxicant concentration (MATC).

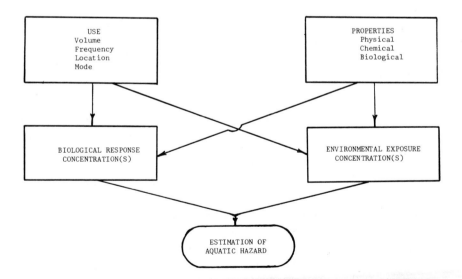

Figure 5-7. Effects of Chemicals on Fish—Factors Involved

This concentration range is classically derived by long-term, chronic bioassays. However, this type of information is often expensive and time-consuming. It is essential to develop screening tests that are scientifically valid, inexpensive, and rapid, which will provide information needed to assess chronic toxicity.

One such screening test currently being studied at Virginia Polytechnic Institute and State University involves the use of fathead minnows, *Pimephales promelas*, ranging in size from 2.5 to 6.0 cm [1 to 2.5 in] as test organisms. The purpose of this project is to develop biomonitoring techniques which detect and predict the concentration of a chemical substance causing impairment in growth and reproductive capabilities upon chronic exposure. The first objective of this study was to develop a system to monitor the activity signal of the fathead minnow. The activity signal differs from the breathing signal explained previously in that it includes fin and body movement as well as opercular movement. The second objective was to determine a parameter of this activity signal which could be used as a short-term indicator of chronic toxicity. The overall hypothesis to be tested is whether a significant, detectable change in the activity signal of the fathead minnow can be used to detect and predict chronic effects of toxicants.

The reasons for working with chronic levels of toxicants already have been explained. The fathead minnow is chosen as the test organism because it has been used universally for chronic bioassays with a background base of chronic toxicity data by which any test system can be compared. There are many problems, however, inherent in dealing with the fathead minnow. Previous biomonitoring projects at Virginia Polytechnic Institute and State University have measured the ventilatory signals of the bluegill, *Lepomis macrochirus* [9, 10]. However, the amount of activity of the fathead minnow and the noise level caused by the degree of amplification required for the smaller ventilatory signals were two major problems not previously encountered when the bluegill was used.

The chamber depicted in figure 5-8 produced the best recordings of electric impulses from the fish of all the chambers tested. The chamber is 30.5 cm [12 in] long and 7.6 cm [3 in] wide, and the shelter is 6.4 cm [2.5 in] long with an inner diameter of 3.8 cm [1.5 in]. A photobeam is directed through the shelter; when this beam is broken, the fish is inside the shelter and the signal picked up by the electrodes is so designated. The electrodes are situated at either end of the shelter, which allows the fish to be monitored as it leaves the shelter. The fish appear calmer when freely allowed to enter and leave the shelter area. This is a flow-through system, and water enters through the baffles, sheets of Plexiglas with small holes, that spread the flow of water and decrease noise caused by the movement of water.

Fifteen of these chambers are placed in an enclosure, and a system delivers toxicant to ten of these chambers (figure 5-9). A peristaltic pump controls

Environmental Contaminants

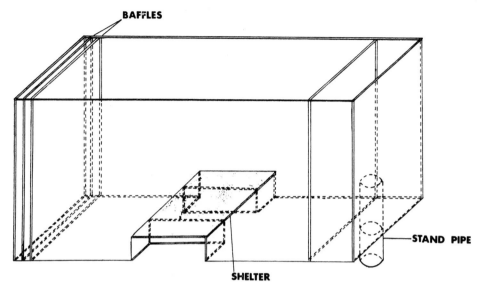

Figure 5-8. Chamber for Exposing Fish and Recording Electric Impulses

the rate of toxicant flow. The fish are exposed to continuous light to encourage them to remain in the shelter where optimum ventilatory signals can be detected.

The electric signal is transmitted from the electrodes to a biopotential amplifier (fifteen amplifiers for the fifteen chambers) [4]. A signal is transmitted from each chamber to the amplifier, which amplifies the signal and filters out background noise (figure 5-10). The PDP8/e minicomputer analyzes the signal and converts it from an analog to a digital form. The computer outputs to a teletype, which prints out opercular rate and other information about the signal at the end of each recording period, and to an oscilloscope, which displays the analog signal of any one of the fifteen fish being tested.

This system is currently being used to determine a parameter of the activity signal which can be used as a short-term indicator of chronic toxicity. Physiological responses (that is, fish opercular rate) have been suggested as such indicators [11]. Much attention has recently been given to the cough response of fish [12, 13]. The high level of activity of the fathead minnow and the difficulty in detecting the ventilatory signal make detection of the cough response impossible. It is even impossible to detect only ventilatory activity; therefore, the activity signal is monitored.

Figure 5-11a depicts a nonstressed signal of the fathead. The fish had been allowed to acclimate to the chamber for three days before the signal was recorded. The bursts of ventilatory activity obvious in this strip-chart recording have also been observed in other nonstressed fish including bluegills.

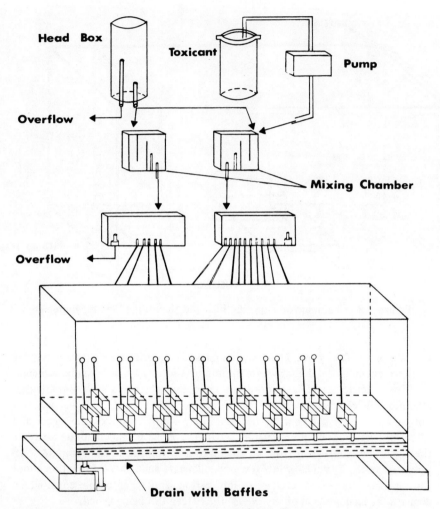

Figure 5-9. Toxicant Delivery Systems

Figure 5-11b represents the signal of the same fish when it has been exposed to 0.6 mg/l zinc. The obvious change in the activity signal suggests several possible indicators such as a change in amplitude, frequency, and disappearance of the burst phenomena.

Several promising parameters have apparently detected the presence of toxicants in preliminary tests. However, more research is needed to determine what would be the best indicator and at what levels. After this system has been tested with zinc, it is proposed to be used with other groups of toxicants with known chronic toxicity data to check for a good predictive correlation.

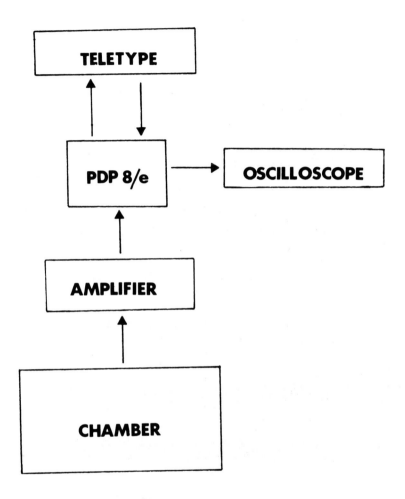

Figure 5-10. System for Sensing and Recording Physiological Response of Fish to Stress

Biomonitoring of Fish Swimming Behavior

In view of the fact that many aquatic organisms are mobile and can modify their behavior to help compensate for certain environmental stresses, a monitor has been developed to analyze changes in the swimming behavior of fish caused by stimulated plumes of toxicants. Originally developed by using a series of

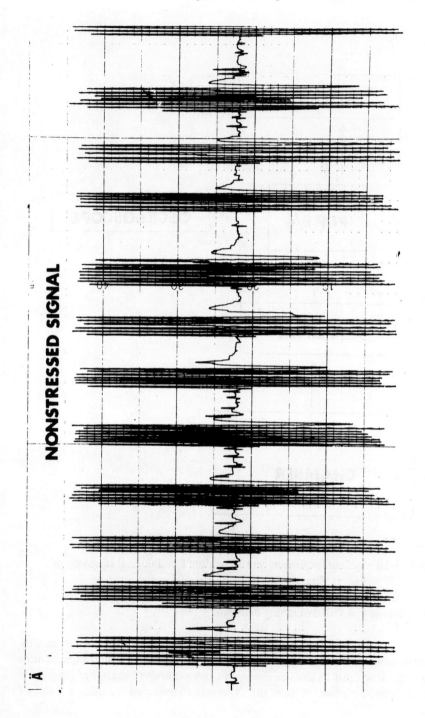

Figure 5-11a. Nonstressed Signals of Fish

Environmental Contaminants

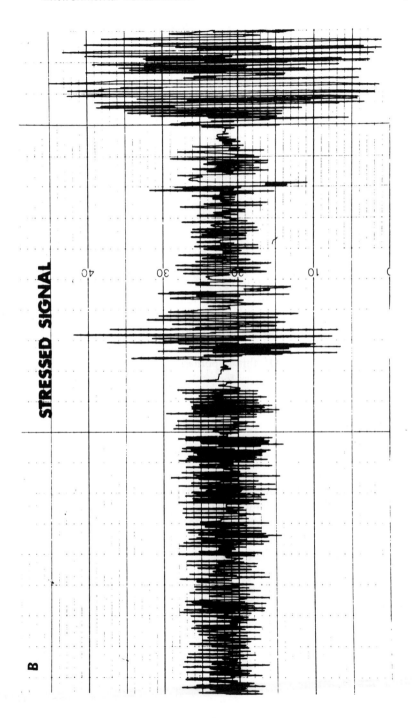

Figure 5-11b. Signals of Exposed Fish

photocells to measure the general activity of fish, the system has been expanded into a video-computer monitor that can analyze the turning and avoidance behavior of fish as well.

The system, outlined in figure 5-12, uses a television camera to track individual test fish in a laminar-flow tank. The camera signal is directed into a video-monitor and a microcomputer which analyzes the position of the fish in an XY grid system at 4-s intervals. Then several behavioral parameters are calculated from these coordinates and are output to a hard copy printed at 10-min intervals. The laminar-flow design of the behavior tank permits a toxicant to be placed into only one side of the tank, creating a steep toxicant gradient across the middle of the observation area. Initial tests with low concentration of ammonium chloride (0.5 mg NH_3-N/L) indicate that bluegills tend to position themselves near the gradient in a type of exploratory behavior. At higher concentrations (5.0 mg NH_3-N/L), both avoidance and preference of the toxic region of the tank have been recorded, responses that seem to be related to the pH of the water and/or the amount of physiological damage sustained by the gill and olfactory tissues. The different types of effects demonstrate the need for recording a variety of behavioral parameters.

Following additional development and testing, this system will be used to test the hypothesis that changes in behavior can be used to predict concentrations of chemical substances causing chronic effects (that is, changes in growth and reproduction). If this hypothesis holds true, the system will be used to screen chemical substances for predicting chronic toxicity at various concentrations.

Conclusions

We feel that the stage is set and the timing is right for biological monitoring to move to the forefront. The capacity currently exists to continually biologically monitor complex industrial and municipal discharges. The system described in this chapter and other similar systems provide us with the means to evaluate the toxic nature and levels of effluents on a real-time basis. Exploratory systems as described in this chapter hold promise of allowing us to predict through short-term experiments the concentrations of a chemical substance causing chronic effects. The challenge to aquatic ecologists is to continue the development of credible test systems to accomplish these goals.

Environmental Contaminants

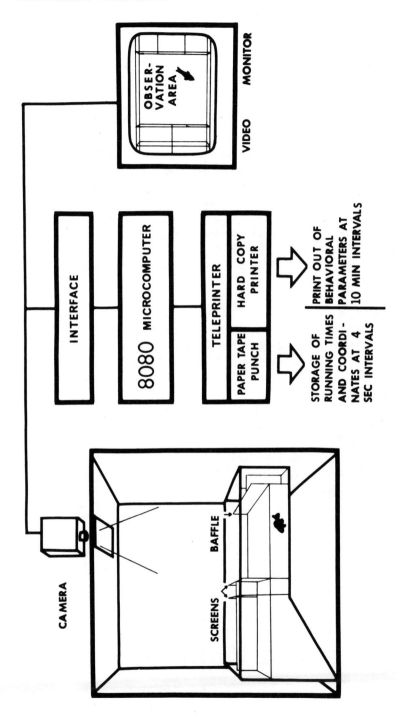

Figure 5-12. TV Tracking—Test Fish

References

[1] Barham, E.G.; Huckabay, W.B.; Gowdy, R.; and Burns, B. Microvolt electric signals from fishes and the environment. *Science* 164:965-968 (1969).

[2] Drummond, R.A., and Dawson, W.F. An inexpensive solid-state amplifier for detecting movements and electrical potentials of fish. *Transactions of the American Fisheries Society* 103(2):391-392 (1974).

[3] Goodman, D., and Weinberger, N.M. Submerged electrodes in an aquarium: Validation of a technique for remote sensing of behavior. *Behavioral Research Methods Instruments* 3(6):281-286 (1971).

[4] Gruber, D.; Cairns, J., Jr.; Dickson, K.L.; Hendricks, A.C.; Hummel, R., III; Maciorowski, A.; van der Schalie, W.H. An inexpensive noise-immune amplifier designed for computer monitoring of ventilatory movements of fish and other biological events. *Transactions of the American Fisheries Society* 106(5):497-499 (1977).

[5] Lonsdale, E.M., and Marshall, W.C., Jr. Noncontact sensing of electric fields surrounding trout. *Instrument Society of America (Transactions)* 12:82-87 (1973).

[6] Spoor, W.A.; Neihersel, T.W.; and Drummond, R.A. An electrode chamber for recording respiratory and other movements of free-swimming animals. *Transactions of the American Fisheries Society* 100(1):22-28 (1971).

[7] Spoor, W.A., and Drummond, R.A. An electrode for detecting movement in gradient tanks. *Transactions of the American Fisheries Society* 101:714-715 (1972).

[8] Hall, J.W.; Arnold, J.C.; Waller, W.T.; and Cairns, J., Jr. A procedure for the detection of pollution by fish movements. *Biometrics* 31(1):11-18 (1975).

[9] Cairns, J., Jr.; and Sparks, R.E. The use of bluegill breathing to detect zinc. Water Pollution Control Research, 1971. Environmental Protection Agency, Ecological Series, Washington, D.C.

[10] Sparks, R.E.; Cairns, J., Jr.; and Heath, A.G. The use of bluegill breathing rates to detect zinc. *Water Research* 6:895-911 (1972).

[11] Spoor, W.A.; Neihersel, T.W.; and Drummond, R.A. An electrode chamber for recording respiratory and other movements of free-swimming animals. *Transactions of the American Fisheries Society* 100(1):22-28 (1971).

[12] Drummond, R.; Olson, G.; and Batterman, A. Cough response and uptake of mercury by brook trout, *Salvelinus fontinalis*, exposed to mercuric compounds at different hydrogen-ion concentrations. *Transactions of the American Fisheries Society* 103(2):244-249 (1974).

[13] Drummond, R.A., and Carlson, R. *Procedures for Measuring Cough (Gill Purge) Rates of Fish*. Washington: Environmental Protection Agency, 1977, EPA 600-3-77-133.

6

Industrial Applications of Biological Monitoring in the Laboratory and Field

Alan W. Maki

Ultimately studies of the effects of chemical substances on aquatic species should contribute in some direct manner to information describing the survival potential of natural surface-water communities potentially coming in contact with dilute concentrations of the test material in their environment. To date, methods for the determination of survival potential of aquatic species during exposures to chemical substance have focused on techniques for describing acute mortality and chronic effects on spawning or reproductive performance. The ecological significance of these methods lies in the direct exposure of the entire life cycle of fish and macroinvertebrates, which may require test periods of one year or more.

Among the more promising short-term predictive approaches has been the biological monitoring of changes in the respiratory activity of fish and aquatic macroinvertebrates as indicators of the organism's response to environmental stress. During this investigation, a system was developed to monitor the ventilation frequency of bluegills, *Lepomis macrochirus*, under continuous-flow exposure to test compounds. The nerve action potential associated with the opening and closing of the operculum was simultaneously monitored with suspended stainless-steel electrodes for sixteen fish, four each from three exposure concentrations and a control.

Test compounds were selected on the basis of previously existing full-life-cycle chronic fish data with which to compare the respiratory data. Good correlations are demonstrated between the chronic maximum acceptable toxicant concentration (MATC) for fathead minnows and the concentrations of test compounds eliciting statistically significant changes in the diurnal ventilation frequencies of exposed bluegills. It appears that the monitoring of aberrant respiratory activity of bluegills for a two-day period has potential predictive value as a scanning tool for a rapid prediction of chronic fish toxicity values.

During a second study, an in-stream biological monitoring program was designed to determine the degradability of an alkyl ethoxylate nonionic surfactant by secondary waste-water treatment and the effects of the waste-water effluent on the benthic fauna of the receiving stream. No increase in surfactant levels in the plant effluent or stream stations was observed during periods when the plant influent was spiked at 5 and 10 mg/l. The acute toxicity of the surfactant is eliminated by secondary waste-water treatment, and the degradation products of an initial 30-mg/l concentration are nontoxic to fathead minnows.

Monitoring and structure analyses of the stream benthic communities indicated significant changes in species composition and density of certain groups of macroinvertebrates during the spiking program. These changes resulted from the natural seasonal life cycles of the species involved and were not brought about by acute or sublethal toxic effects of either parent nonionic or degradation products.

These two studies indicate the value of biological monitoring programs for the establishment of ecologically meaningful water quality critera and the application of biological monitoring techniques to examine the structure and function of aquatic communities potentially exposed to new chemical substances.

Laboratory Biological Monitoring

Changes in the respiratory activity of fish and aquatic macroinvertebrates have been used by several investigators as indicators of an aquatic organism's response to environmental stress [1, 2, 3, 4]. Although these investigations have proved very useful in describing sublethal responses with direct implications toward mode-of-action studies, the interpretation of the ecological significance of these numerous respiratory responses remains difficult.

The present investigation was designed to (1) develop a method to continuously monitor the ventilation frequency of bluegills under continuous-flow exposures to easily biodegradable test substances, accounting for intrinsic and diurnal variations in normal activity, and (2) examine correlations between the observed no-effect concentrations determined by this method and previously existing full-life-cycle chronic effect data for the fathead minnow, *Pimephales promelas*.

Methods and Materials

Description of Test System

Test fish were enclosed within 4 cm X 11.5 cm X 10 cm deep individual glass compartments painted flat black on the outside and provided with single-pole, stainless-steel electrodes at anterior and posterior ends. A total of three treatments and a control, all containing four replicate chambers, were simultaneously supplied with test material concentrations and dilution water from a modified 0.5-liter proportional diluter [5]. A four-way flow-splitting chamber ensured a flow rate of 125 ± 10 ml/min through each individual compartment (figure 6-1).

All tests were conducted at 21 ± 2°C under a 12-hr photoperiod. Incandescent lamps were controlled by a solid-state dimming device to gradually intensify over a 30-min period during morning hours and to gradually dim

Industrial Applications

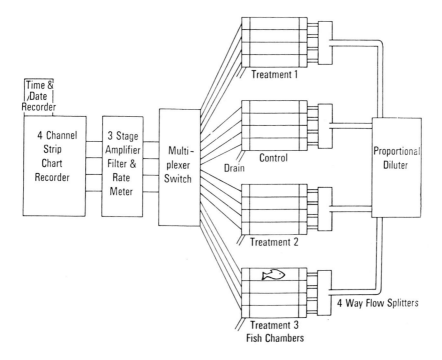

Figure 6-1. Test System Used for All Ventilation Frequency Tests with Bluegills Showing the Three Treatments and Control Chamber Arrays, Switching Device, Amplifier, and Four-Channel Recorder

over the same period during the evening. All fish compartments were placed within a 1.5 m × 1.5 m tray and screened from outside movements with an opaque black curtain. Dilution water used for all investigations was filtered well water, and the test materials used are characterized in table 6-1. Generic names of test materials are used throughout the text.

Test Procedure

Juvenile bluegills of 8 ± 1.5 cm were obtained from Fender's Fish Hatchery, Baltic, Ohio. Fish were held at 13°C in 540-liter recirculating tanks provided with a continuous input of filtered well water. Test fish were acclimated to test conditions over a two-week period in separate acclimation tanks. Fish were placed within the test chambers and allowed to acclimate to the chamber for a five-day period prior to any recordings. An experiment was then initiated by recording individual fish activities over a three-day control period. Treatment

Table 6-1
Chemical Characterization of the Test Surfactants and Single Builder

Generic Name	Structure	Chemical Characterization
$C_{11.8}$ Linear alkyl benzene sulfonate	$CH_3-(CH_2)_x-CH_3$ with $SO_3^--Na^+$ on benzene ring	Linear alkyl benzene sulfonate, anionic, mean alkyl chain length = 11.8; range C_{10-14}; mean phenyl position = 3.76; mean molecular weight = 345
C_{13} Linear alkyl benzene sulfonate		Anionic, mean alkyl chain length = 13.3, range C_{10-14}; mean phenyl position = 4.5; mean molecular weight = 367
Alkyl ethoxylate sulfate	$CH_3(CH_2)_{16}-O(C_2H_4O)_3OSO_3^--Na^+$	Anionic; molecular weight = 490
$C_{14.5}$ alkyl ethoxylate	$CH_3-(CH_2)_x-(C_2H_2O)_yH$	Nonionic x = 7 to 19, mean = 13.5 y = 0 to 12, mean = 7.0
$C_{12.5}$ alkyl ethoxylate		Nonionic x = 7 to 19, mean = 11.5 y = 0 to 12, mean = 6.5
Amine oxide	$CH_3-(CH_2)_x-N^+(CH_3)_2-O^-$	Alkyl dimethyl amine oxide; nonionic; molecular weight = 243; x = mean 11.7
Trisodium nitrilotriacetate	$Na-O-C(=O)-CH_2-N(CH_2-C(=O)-O-Na)_2$	Trisodium nitrilotriacetate; builder; molecular weight = 257

was initiated by connecting a Mariotte bottle filled with stock concentrations of test material to the diluter. No solvents were used since the test surfactants are highly water-soluble. Exposure continued for a two-day period.

Statistical Analysis

The entire strip-chart record of the three-day control period and the two-day exposure was counted manually for ventilation rates only. Original objectives called for examination of cough frequency; however, significant differences in interpretation of cough frequency occasionally greater than 100 percent between personnel counting the same chart record made absolute counts highly subjective and resulted in the focus on ventilation rates, where excellent agreement between counting personnel was obtained. For purposes of statistical analysis, the day was lumped into four time periods: period 1: midnight to 6:00 a.m.; period 2: 6:00 a.m. to noon; period 3: noon to 6:00 p.m.; period 4: 6:00 p.m. to midnight. Responses recorded at 2-hr intervals were placed into these periods and presented as a mean rate, with upper and lower 95 percent confidence limits calculated for each period. The analysis of variance for each data set was done with a computerized package, PROC ANOVA [6]. Differences in rates among treatment groups were examined during the pretreatment control period to ascertain similarity of responses, and subsequently treatment groups were compared to the control group during exposure to examine dose-related effects. Significance of interaction effects was examined for the following: fish X group, groups X period, and period X fish (group). Dunnett's procedure was then used to compare mean ventilation rates from each treatment concentration to the simultaneously determined mean rates for control fish [7]. No statistically significant difference between control and treatment means defined the no-effect concentration for each test material.

Results

Ventilation Frequency

The effects of a test substance can be discerned over the normal diurnal activity pattern of bluegills (figure 6-2). A typical example of the data produced by this method for amine oxide demonstrates a statistically significant increase in the mean ventilation frequency over all four time periods of the day.

An examination of the responses recorded for the two nonionic surfactants, C_{14-15} and C_{12-13} alkyl ethoxylates, indicates that ventilation rates do not increase during exposure to these materials; instead they are found to decrease at the concentrations tested. Comparisons for all four time periods and across

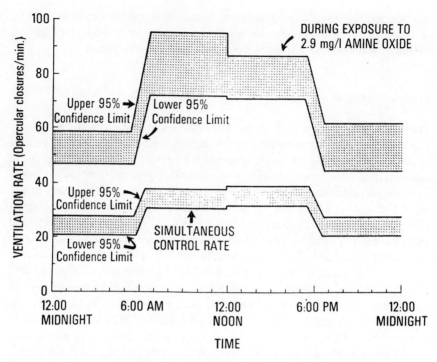

Figure 6-2. Upper and Lower 95 Percent Confidence Intervals for Bluegill Ventilation Rate during Exposure to 2.9 ml/l Amine Oxide Compared to Simultaneously Recorded Control Rates

the concentrations tested indicate that the C_{12-13} alkyl ethoxylate acts as a somewhat greater suppressant on ventilation rates. An examination of the responses recorded for the two nonionic surfactants, C_{14-15} and C_{12-13} alkyl ethoxylates, indicates that ventilation rates do not increase during exposure to these materials; instead, they are found to decrease at the concentrations tested. Comparisons for all four time periods and across the concentrations tested indicate that the C_{12-13} alkyl ethoxylate acts as a somewhat greater suppressant on ventilation rates than does the C_{14-15} chain length. The recorded rates under exposure to 1.2 mg/l of C_{12-13} alkyl ethoxylate are significantly different from simultaneously recorded control rates for unexposed fish during all four time periods (figure 6-3). These data indicate that ventilation rates were consistently reduced by 30 to 50 percent at this test concentration.

Statistical Analysis

The observed ventilation rates during exposure to test materials were averaged means to ascertain statistical significance of effects by using Dunnett's procedure

Industrial Applications

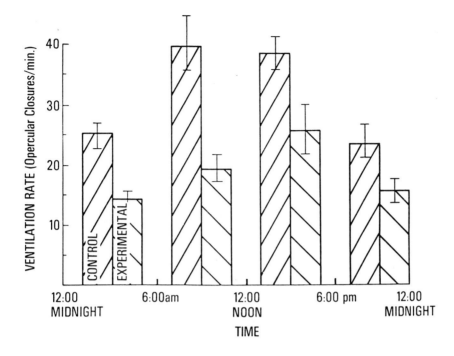

Figure 6-3. Decrease in Ventilation Rates of Bluegills during Exposure to 1.2 mg/l C_{12-13} Alkyl Ethoxylate as Compared to Simultaneously Recorded Control Rates

(table 6-2). This procedure allowed a statistical determination of a no-effect concentration of each test material by declaring observed rates during exposure to certain concentrations as indistinguishable from control rates. For $C_{11.8}$ Linear alkyl benzene sulfonate (LAS), only the highest treatment concentration of 2.19 mg/l is significant, while mean rates for the lower test concentrations are slightly elevated over pretreatment rates to a statistically insignificant degree. Postexposure rates for this compound indicate a return to pretreatment control rates within a 48-hr period following exposure. Although the pretreatment control period indicates no difference between controls and all three test groups for C_{13} LAS during the preexposure period, the rates recorded during exposure periods differ significantly for all concentrations tested. Similarly, with all test materials no significant difference is shown between controls or treatment concentration means during the pretreatment control period, indicating all fish were responding similarly prior to initiation of the test concentrations. Two test materials, C_{14-15} alkyl ethoxylate and NTA, indicate slight changes in mean ventilation rates during exposure periods; however, they prove to be not significant, thus indicating a no-effect concentration in excess of the highest dose levels for these materials. Data for these remaining three test materials (alkyl

Table 6-2
Summary of Dunnett's Procedure for Comparison of Treatment Group Means to Simultaneous Control Mean

Test Material	Test Concentration	24-hr Mean Ventilation Rates (Opercular Closures/min)			Test Material	Test Concentration		
		Preexposure Control	During Exposure	Post-Exposure			Preexposure Control	During Exposure
$C_{11.8}$ LAS	Control	23.7	25.4	26.1	C_{12-13} alkyl ethoxylate	Control	32.3	40.8
	0.45	21.1	27.3	26.4		0.26	27.1	32.9
	1.06	20.1	27.8	24.1		0.54	28.4	29.8
	2.19	22.8	33.1*	25.0		1.20	32.8	19.8
	D statistic	n.s. 4.7	4.7	n.s. 4.5		D statistic	n.s. 8.0	8.1
C_{13} LAS	Control	26.8	32.5		Amine oxide	Control	29.5	30.9
	0.39	27.6	57.5*			0.93	34.5	42.7
	0.77	28.2	70.1*			1.39	30.5	36.4
	1.50	22.9	107.7*			2.99	26.4	69.3
	D statistic	n.s. 6.4	12.2			D statistic	n.s. 5.6	12.2
Alkyl ethoxylate sulfate	Control	23.0	26.4		NTA	Control	27.6	31.0
	0.24	22.4	30.1			37.7	27.3	33.4
	0.39	24.0	38.8*			70.9	28.3	34.8
	0.58	24.6	64.2*			181.2	31.8	36.1
	D statistic	n.s. 5.7	11.3			D statistic	n.s. 4.5	n.s. 5.4
C_{14}–C_{15} alkyl ethoxylate	Control	23.8	31.7	31.1				
	0.56	28.5	32.0	30.5				
	1.08	28.8	28.4	30.5				
	1.56	24.7	27.9	31.7				
	D statistic	n.s. 6.9	n.s. 5.4	n.s. 8.3				

Indicated treatment means are significantly different from control means since they differ from control means at least by the value indicated by individual D statistics.
*Level of significance for $p = 0.01$.
n.s.: no significant difference.

ethoxylate sulfate, C_{12-13} alkyl ethoxylate, and amine oxide) indicate at least the highest exposure concentration to be significantly different from control rates. Also, data for the lowest test concentration for these materials prove to be indistinguishable from control rates, thus defining a no-discernible-effect concentration (table 6-2).

Discussion

Predictive Utility

A comparison of the statistically determined no-effect concentrations based on ventilation rate data for several surfactants demonstrates good agreement with known MATC values determined from actual chronic tests with the fathead minnow, *Pimephales promelas* (table 6-3). The values in both columns are expressed as a range, with the lower value being the actually measured test concentration causing no effects. The upper value is the lowest of the high-level no-effect concentrations and therefore it lies between the two measured concentrations. Of the surfactants tested, only the data for C_{14-15} alkyl ethoxylate fail to agree reasonably with the MATC value, indicating the technique may have limited value as a predictive tool for this determination of chronically toxic concentrations of some nonionic surfactants. For the remaining surfactants and single builder, the agreement between test endpoints from the five-day ventilation frequency test and the chronic MATC value is reasonably accurate, to underscore the predictive utility of the ventilation test as a tool for the early estimation of chronic fish toxicity.

For the surfactant materials examined, which are all highly biodegradable, the method is predictive of the known MATC values for the intact parent molecule in the absence of any significant biodegradation [8]. In the present context, the method does not yield data for the assessment of effects of actual environmental forms of these test materials since significant environmental biodegradation is known to substantially reduce the acute and chronic toxicity of the parent molecule. The cost and time savings of the ventilation test present an opportunity to develop an estimate of chronic toxicity early in a hazard evaluation program designed for risk assessment of new materials being considered for commercialization, and this test may provide a tool for rapid estimation of chronic toxicity of many more materials for which previous full-life-cycle testing was economically unjustifiable.

As methods for remote sensing and biomonitoring of test material effects on the behavior of aquatic organisms proliferate, some attempts must be made to standardize apparatus and methods to ensure that a comparable and relevant data base can be developed to properly assess the significance and application to receiving water population.

Table 6-3
Comparison of Chronic MATC Values for Fathead Minnow and the No-Effect Concentrations Shown in the Five-Day Ventilation Rate Tests with Bluegills

Test Material	Fathead Minnow MATC[a] (mg/l)	Ventilation Rate No Effect (mg/l)
$C_{11.8}$ LAS	$0.70 < X < 1.1$	$1.06 < X < 2.19$
C_{13} LAS	$0.10 < X < 0.24$	$X < 0.39$
Alkyl ethoxylate sulfate	$0.13 < X < 0.20$	$0.24 < X < 0.39$
C_{14}–C_{15} alkyl ethoxylate	$0.21 < X < 0.28$	$X > 1.56$
C_{12}–C_{13} alkyl ethoxylate	$0.32 < X < 1.0$	$0.26 < X < 0.54$
Amine oxide	$0.50 < X < 1.0$	$1.39 < X < 2.99$
Nitrilotriacetate	$X > 54$[b]	$X > 181$

[a]Unpublished data.

[b]A.W. Maki, K.W. Stewart, and J.K.G. Silvey, The effects of dibrom on respiratory activity, *Transactions of the American Fisheries Society* 102:806-815 (1973).

Field Biological Monitoring

Introduction

The assessment of hazard to aquatic life associated with the use of a particular chemical necessarily relies heavily on the predictive value of an assortment of controlled laboratory tests. The investigator then takes into account the observed physical, chemical, and biological effects of the material in determining the probability of creating measurable, real-world environmental effects associated with intended use of the new substance. In this study an alkyl ethoxylate nonionic surfactant was continuously introduced into the influent of a wastewater treatment plant at exaggerated concentrations to provide environmental data under field conditions to evaluate its removability and toxicity to aquatic life.

This study employed an integration of qualitative and quantitative sampling techniques with short-term acute toxicity tests to characterize the effects of an activated sludge effluent on the benthic community existing within a receiving stream. Specific objectives of the study were:

1. Develop a quantitative and qualitative description of the macroinvertebrate fauna existing above and at two stations below the Reynoldsburg sewage treatment plant on Blacklick Creek.
2. Quantify specific changes beyond the natural seasonal emergence patterns that may occur within these macroinvertebrate communities during and

following a nonionic surfactant spiking program at the sewage treatment plant that may be attributable to either direct toxicity from untreated surfactant or residual toxicity of nonionic degradation products and metabolites.
3. Determine the acute fish toxicity of the undergraded and biologically degraded surfactant as well as the rate of loss of toxicity during biodegradation through the conduct of effluent toxicity tests with the fathead minnow.

Methods and Materials

The study was conducted in two phases during 1974-1975 (table 6-4). The facility chosen for field testing was the Reynoldsburg sewage treatment plant, a contact-stabilization plant with a design capacity of 7,600 m^3/day [2.0 MG/day] located in Reynoldsburg, Ohio. The added amounts of Neodol 45-7 used in this investigation bring the influent concentrations up to a point well above those typically encountered in municipal influents.

Fish used in all toxicity tests were fathead minnows, *Pimephales promelas* (Rafinesque), obtained from a commercial hatchery in Baltic, Ohio. Fish were acclimated to test temperature, and the toxicity tests were conducted following recommended guidelines [9].

Stream Survey. The Reynoldsburg treatment plant is located on the eastern bank of Blacklick Creek, a tributary to the Scioto River. Three stations were selected on Blacklick Creek to bracket the point of sewage outfall and are referred to throughout as follows: *upstream*, the first location selected in a riffle area approximately 100m upstream from the effluent outfall; *downstream*,

Table 6-4
Neodol 45-7 Treatment Periods and Test Concentrations during Summer- and Winter-Phase Studies at the Reynoldsburg Sewage Treatment Plant[a]

Experimental Period	Summer Phase	Winter Phase
1. Pretreatment background	July 9 to August 25	January 13 to February 2
2. 5-gm/l treatment 8 a.m. to 10 p.m. 10 p.m. to 8 a.m.	August 26 to September 8 25 kg/day (5.3 mg/l) 25 kg/day (5.3 mg/l)	February 3 to February 16 30 kg/day (6.4 mg/l) 30 kg/day (6.4 mg/l)
3. 10-mg/l treatment 8 a.m. to 10 p.m. 10 p.m. to 8 a.m.	September 9 to October 27 66 kg/day (14 mg/l) 37 kg/day (8 mg/l)	February 17 to March 30 86 kg/day (12 mg/l) 42 kg/day (6 mg/l)
4. Posttreatment	October 28 to November 8	March 31 to April 11

[a]Differential dosage rates reflect seasonal and diurnal flow variations.

the second location in a riffle area 100 m downstream from the effluent outfall; *a recovery zone*, approximately 3.2 km below the effluent outfall. These three sampling areas were selected for their physical environmental similarities. Each was located within a wooded area with a fully forested canopy with similar flow patterns and current regime. Water depth varied from 6-cm to 30-cm diameter at each station.

Six samples were taken at each of the three stations during each sample period. Sampling was done by using a specially modified Surber square foot sampler with a 20-cm foam-padded bottom to conform to bottom irregularities, a screened front to allow water current to wash entrained fauna into the removable net. This sampler ensured that 6 ft^2 of stream bottom was uniformly and efficiently sampled during each sampling period.

Macroinvertebrates and incidentally entrapped fish fauna were picked free of the samples net, preserved in 1096 formalin, and returned to the laboratory for analysis. All individuals were hand-separated from the detrital mass collected during the sampling process, placed into separate vials by taxonomic groups, and identified to genus or species when possible.

All sample data were normalized to represent total standing crop of individuals per square foot of stream bottom. The Shannon-Weiner diversity index, as described by Wilhm and Dorris, was calculated by employing sample data to estimate the actual population diversity [10]. The index of species equitability E was calculated after MacArthur [11]. This measurement serves as a descriptor of the relative evenness of species distribution, with generally higher values indicating fewer species, all found equally distributed, and lower values indicating unequal distributions, some common and some rare individuals.

Water Chemistry. Water samples were taken from all three stations each weekday throughout the June to October sampling period. Simultaneously, readings of dissolved oxygen and water temperature were also taken. All water samples were preserved with 1 percent formalin solution and analyzed for surfactant concentrations. Since no specific analytical method is available for Neodol 45-7, a modified cobalt-thiocyanate (CTAS) method was used to estimate the nonionic surfactant concentrations present at all three stations [12]. It should be emphasized that CTAS levels do not represent only nonionic and synthetic organic substances yielding positive CTAS activity. Anionic surfactant was analyzed by a methylene blue active substance (MBAS) procedure with special modifications employing an HCl hydrolysis and ethyl ether extraction to minimize interferences [13].

Results

Toxicity of Degradation Intermediates. Surfactant die-away tests in stream water and secondary effluent were conducted with initial concentrations of 3.0 and 10 mg/l of the surfactant (figure 6-4). Fish introduced into stream

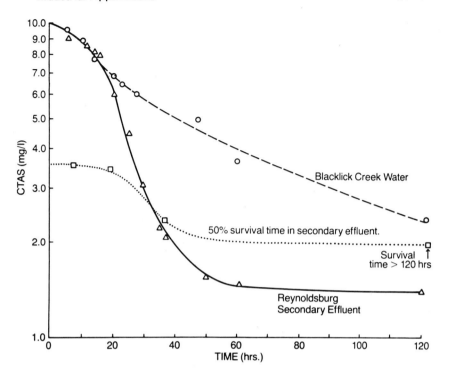

Figure 6-4. Biodegradation Curves for Neodol 45-7 Solutions in Secondary Effluent and Blacklick Creek Water, Superimposed with Fathead Minnow 50 Percent Survival Time Data, Demonstrating Loss of Toxicity with Loss of CTAS Response

water at the same time as the surfactant (0 hr) died within 12 hr. Six hours after the surfactant was added, the measured CTAS level was 2.38 mg/l and fish introduced at this time died within the following 28 hr. All fish introduced 24 hr after the surfactant was added survived. When secondary effluent was used, fish introduced at 0 hr all died within 12 hr. Fish introduced at 6 and 12 hr, at remaining CTAS levels of 2.7 and 2.4 mg/l, respectively, experienced 70 percent mortality within the following 18 hr with no additional deaths being observed at 24 hr. Fish exposed after 24 hr survived the duration of the test.

Macroinvertebrate Community Structure. The numbers of bottom fauna taxa per square foot of stream bottom during the experimental periods are summarized in a histogram, figure 6-5*a*. The highest mean number of taxa and the highest variability were found in the recovery zone station where a mean of 8.8 genera was found during the 5-mg/l treatment period and a high of 13.3 genera during the posttreatment period. The number of taxa found at the upstream stations was consistently lower than at the recovery area. Data here ranged

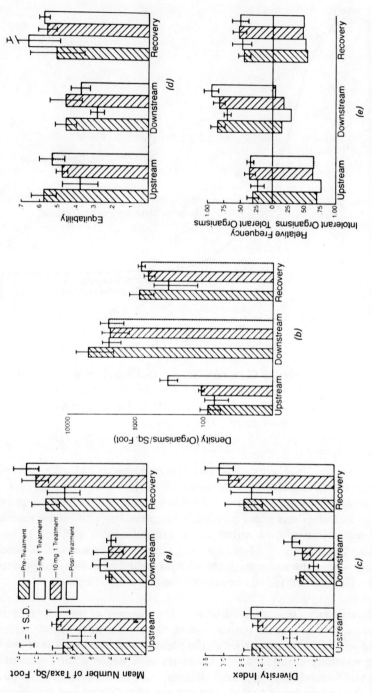

Figure 6-5. Characterization of the Macroinvertebrate Communities Existing in Blacklick Creek During the Neodol 45-7 Spiking Program at the Reynoldsburg Waste-water Treatment Plant

from a mean low of 7.1 during the 5-mg/l spiking period to 9.8 genera per square foot during the 10-mg/l period.

The lower number of taxa observed at recovery and upstream areas during the 5-mg/l spiking period is attributed to a period of heavy late-summer showers which at times raised the stream level 0.3 to 0.6 m above levels observed during other experimental periods. The inverse effect was observed at the downstream station where the highest number of taxa, 5.5 per square foot, was found during the 5-mg/l spiking period. Although all other periods demonstrate a mean of about 4 genera per station, severely depressed because of the influence of the treatment plant effluent, the increase in genera during the 5-mg/l spiking period is explained by the fact that high water flows effectively scoured individuals from the upstream area and redeposited them in the downstream sampling area. Only during this sampling period were Psephenid and Elmid beetles found in the downstream area. They did not naturally exist there, and the lack of their presence in the successive sampling intervals indicates that they disappeared from that area rapidly through continued downstream drift. A profound difference exists between the mean density of individuals per square foot found at each sampling area (figure 6-5b). The highest densities, ranging from 2,300 to 4,800 individuals per square foot, were consistently found at the downstream sampling area. The high numbers sampled during the August pretreatment period represent a dense population of midges, *Chironomus* spp., which emerged in early September. Sampling throughout August demonstrated that a high proportion of the midge population in this area appeared to be growing synchronously, as evidenced by similarity of larval size, and hence would be expected to emerge simultaneously. Evidence for the natural emergence included high numbers of pupal exuviae in backwaters and numerous winged adults during this period. At any rate, the standard deviations for all four experimental periods overlap and demonstrate no variations attributable to the treatment periods.

The intermediate density of organisms was consistently found at the recovery zone. Here, the mean variation ranged from 320 to 860 organisms per square foot during the sampling program. A different population of filter-feeding blackfly larvae was the predominant species during August. Their emergence late in the month, along with the scouring effect and subsequent drift of individuals during the period of high water flows, caused the lower density observed during the 5-mg/l spiking period in early September. The successive increases in density observed throughout the 10-mg/l and posttreatment periods of October and November represent recolonization and recruitment of univoltive caddisfly species to a size large enough to be retained by the sampler net. These eggs then hatched to early instar larvae which, although present throughout the summer months, were too small to be retained in the sampler net. As these larva grew through successive instars during the early fall, they began to be retained in increasing numbers, resulting in the higher mean density recorded for this station.

The information from these two previous analyses of number of taxa and density is instrumental in the development of diversity indices for each station and experimental period (figure 6-5c). The highest mean diversity value of 3.12 was obtained in the recovery zone during late October. The lowest mean value observed at this station in September was 2.25. The relatively wide standard deviations found here indicate the variability of the community compared to upstream and downstream areas. This area is by no means free of the influence of the treatment plant, even though the diversity figures, when considered alone, compare favorably and occasionally exceed the upstream station. The mean diversity value for the 5-mg/l period reflects the period of high water flows, and the successively increasing values for the 10-mg/l and posttreatment periods indicate the recruitment and growth of new individuals discussed earlier.

Diversity indices for the upstream location ranged from 1.35 to 2.40, with the low value again attributable to high water flows and no significant differences between the other experimental periods. The high density of individuals and relatively low number of taxa recorded for the downstream station combine to generate the significantly lower diversity values observed at this station throughout the sampling program. The emergence of a large segment of the midge population in late August brought about the lowest mean diversity figure of 0.61 during the 5-mg/l spiking period. The figures for the remaining experimental periods are relatively consistent and vary between 0.85 and 1.20, demonstrating no consistent change which can be correlated to the plant treatment periods.

Calculations of community equitability for the three sampling stations varied from a low of 0.29 at the downstream location to a maximum of 0.66 in the recovery zone, with both extremes occurring during the period of high water during the 5-mg/l dose (figure 6-5d). With this period as the exception, the equitability values for the remaining periods are consistent within each station and demonstrate no significant variations correlated with the plant treatment periods.

The species collected from the three sampling stations on Blacklick Creek during each experimental period were then tabulated and compared with published listings of indicator organisms [14, 15]. A good comparison was obtained for the community structure of pollution-tolerant and -intolerant species (figure 6-5e). The upstream area shows the highest percentage of intolerant species ranging from 65 to 75 percent of total fauna. The impact of the treatment plant effluent is evident at the downstream station, where the percentage of intolerant organisms drops to 6 to 28 percent. The higher figure was obtained during the period of high water flow when upstream intolerant species had drifted downstream to this location. Generally, throughout the experimental period the figures ranged between 6 and 12 percent. An approximately equal distribution of 50 percent tolerant and intolerant species was observed in the recovery zone, which further indicates that this area was affected by the plant effluent considerably before the nonionic treatment program of the Reynoldsburg plant.

During the process of placing the modified Surber sampling in position on the appropriate stream transect, occasional fish species were entrained within the net. Although in no way meant to be quantitative, these incidental fish collections did serve as a qualitative indication of existing water quality. Seven species of fish were sampled, five species within the darter family and two minnow species. The major significance lies in the fact that although twenty-seven and twenty-nine individuals were collected from upstream and recovery zone areas, respectively, at no time were any fish collected from the downstream station.

Conclusions

1. No increase in nonionic surfactant levels as measured by CTAS in the plant effluent, downstream, or recovery zone stations was observed during periods when the influent of the Reynoldsburg sewage treatment plant was spiked at concentrations as high as 5 and 10 mg/l.
2. The 96-hr static LC50 for fathead minnows exposed to Neodol 45-7 in carbon-filtered tap water is 1.2 mg/l; in Blacklick Creek water it is 1.38 mg/l; and in secondary effluent it is 2.48 mg/l.
3. The acute toxicity of Neodol 45-7 is completely eliminated by secondary waste-water treatment at the Reynoldsburg plant. The acute toxicity of the surfactant is also lost, but at a slightly slower rate in the relatively bacteriafree natural surface water taken from Blacklick Creek.
4. The degradation products of Neodol 45-7 are apparently nontoxic, based on 96-hr acute exposures of fathead minnows, even when the surfactant was originally present at a concentration of 30 mg/l.
5. Significant differences in macroinvertebrate community structure (as measured by density of individuals and number of taxa per unit area, diversity indices, equitability function, and indicator organisms) existed among the upstream, downstream, and recovery zone areas prior to the spiking program at the treatment plant. These community differences represent a classic example of a stream receiving an organic waste high in biological oxygen demand (BOD).
6. Structure analyses of the benthic communities conducted during the pretreatment, 5-mg/l spiking, 10-mg/l spiking, and posttreatment periods indicated significant changes in species composition and density of certain groups of macroinvertebrates during the spiking program. These changes resulted from the natural seasonal life cycles of the species involved and were not brought about by acute or sublethal toxic effects of either parent nonionic or degradation products.

References

[1] Belding, D.L. The respiratory movements of fish as an indicator of the toxic environment. *Transactions of the American Fisheries Society* 59:238-246 (1929).

[2] Carpenter, K.E. Further researches on the action of metallic salts on fishes. *Journal of Experimental Zoology* 56:407-422 (1930).

[3] Knight, A.W. and Gaufin, A.R. Relative importance of varying oxygen concentration, temperature and water flow on the mechanical activity and survival of *plecoptera nymph*. *Proceedings of the Utah Academy of Sciences, Arts and Letters* 41:14-28 (1964).

[4] Maki, A.W., Stewart, K.W., and Silvey, J.K.G., The effects of dibrom on respiratory activity of the stonefly, *hydroperla crosby*, hellgrammite, and the golden shiner, notemigonus crysoleucas. *Transactions of the American Fisheries Society* 102:806-815 (1973).

[5] Mount, D.I., and Brungs, W.A. A simplified dosing apparatus for fish toxicology studies. *Water Research* 1:21-29 (1967).

[6] Barr, A.J.; Goodnight, J.H.; Sall, U.P.; and Helwig, J.T. *Statistical Analysis System*. Raleigh, N.C.: Sparks Press, 1967.

[7] Steel, R.G.D., and Torrie, J.H. *Principles and Procedures of Statistics.* New York: McGraw-Hill, 1960.

[8] Swisher, R.D. *Surfactant Biodegradation.* New York: Marcel Dekker, Inc., 1970.

[9] "Methods for Acute Toxicity Tests with Fish, Macroinvertebrates and Amphibians." Washington: Environmental Protection Agency, 1975, EPA 660/3-75-009.

[10] Wilhm, J.L., and Dorris, T.C. Biological parameters for water quality control. *Bioscience* 18:477 (1968).

[11] MacArthur, R.H. Patterns of species diversity. *Biological Review of the Cambridge Philosophical Society (G.B.)* 40:510 (1965).

[12] Boyer, S.L.; Guin, K.R.; Kelley, R.M.; Mausner, M.L.; Robinson, H.F.; Schmitt, T.M.; Stahl, L.R.; and Stezkorn, E.A. Analytical methodology for non-ionic surfactants. *Environmental Science and Technology* 11(13): 1167-1171 (1977).

[13] American Public Health Association. *Standard Methods for the Examination of Water and Wastewater*, 13th ed. Washington, 1971.

[14] Mason, W.F., Jr.; Lewis, P.A.; and Anderson, J.B. "Macroinvertebrate Collections and Water Quality Monitoring in the Ohio River Basin, 1963-1967." Cooperative Report, Office of Technical Programs, EPA. Cincinnati, Ohio, 1971.

[51] Weber, C.I. "Biological Field and Laboratory Methods for Measuring the Quality of Surface Waters and Effluents." Cincinnati, Ohio: Office of Research and Development, EPA, 1973.

7

Biomonitoring of Coastal Waters— An Overview

John P. Couch, Frank G. Lowman, and Ford A. Cross

Estuarine and coastal waters of the United States not only serve as prime habitat for a significant fraction of commercial and recreational marine fisheries, but also receive industrial and municipal wastes from a rapidly expanding coastal economy. In order to ensure that levels of contaminants in coastal ecosystems do not affect either public health or fisheries resources adversely, both state and federal agencies are conducting a variety of biomonitoring programs. In addition, generic research projects are underway at several university, state, and federal laboratories to determine which environmental and physiological factors regulate the body burden of contaminants in marine organisms. This information can then be used to help interpret data obtained in biomonitoring programs and to allow more precise predictions of contaminant levels in biota prior to discharge.

At present, the Environmental Protection Agency is in the process of establishing two monitoring programs to determine the consequences of anthropogenic inputs of contaminants into estuaries and coastal waters. One of these programs is known as the "Mussel Watch Network." This is a program being administered by the EPA Environmental Research Laboratory of Narragansett, Rhode Island. Coordination of this program is being undertaken by the Scripps Institute of Oceanography, University of California, San Diego under the direction of Dr. Goldberg. Five marine laboratories are participating in this effort [1, 2].

The Mussel Watch concept was evolved in 1976 as a means for monitoring trace pollutants in seawater or concentrations ranging from parts per quadrillion (10^{-15} g/g) to parts per trillion (10^{-12} g/g). The species of bivalves selected as surveillance organisms were mussels (*Mytilus*) and oysters (*Ostrea* or *Crossostrea*). The ability of these bivalves to concentrate pollutants well above environmental levels eliminates the need to extract and measure them from large volumes of seawater by expensive and time-consuming physicochemical means. These bivalves perform a key role in separating substances such as heavy metals, transuranic elements, petroleum, and halogenated hydrocarbons to enable their identification and presence in seawater. [4]

Variations and discrepancies in data resulting from species, sex, and age differences in these organisms are also being evaluated for their significance.

The carcinogen research program at the EPA Gulf Breeze Laboratory is concerned with the development and use of select estuarine and marine species as monitoring and indicator organisms for the detection of carcinogens. The following three objectives indicate approaches to implementing the program.

1. Carcinogenic or suspect carcinogenic substances enter the aquatic environment as pollutants and pose a multithreat to aquatic ecosystems. Because most carcinogens are mutagens, the major risks to aquatic ecosystems and component species are long-term mutagenic and teratogenic effects expressed at both organism and population levels. Populations of valuable aquatic species impacted by mutagenic substances will be studied in order to predict effects on population stability and survival. Commercially and ecologically valuable species such as fishes, oysters, and shrimps may be adversely affected by carcinogens in the form of cellular proliferative diseases and mutagenic effects.

2. The intake of carcinogens, mutagens, and teratogens by humans comes in part through food. Our chief concern in this area is the need to determine if aquatic species accumulate and convert procarcinogens to proximal carcinogens and thus pose a direct carcinogenic threat to humans in their consumption of seafood. Therefore it is essential to know the routes, rates, and reservoirs involved in the accumulation of these compounds in the marine food web.

3. Aquatic organisms accumulate carcinogens from runoff, fallout, and discharge of pollutants into the aquatic portion of the biosphere which behaves as the ultimate pollutant "sink." Select species in the aquatic environment have potential to be used as sentinel or indicator systems to reflect the presence, behavior, and effects of carcinogens, mutagens, and teratogens. Laboratory and field studies of select species for uptake, accumulation, and effects of known carcinogens should reveal the better modes for utilization of aquatic species as indicators. Results of studies with sentinel species can be used to determine when further testing is required and may offer pertinent information on mechanisms of effects.

To date, several experimental studies have been completed on uptake, accumulation, and effects of known mammalian carcinogens [for example, benzo (*a*)-pyrene and 3-methylcholanthrene] in oysters [3].

Methods

This study will be essentially an epizootiological and chemical analytical survey. Five or six regional stations will be sampled for fish, shellfish, water, and sediment samples. Major sampling stations are located at Apalachicola, Florida. Each substation will be sampled once a month for 24 to 36 months. The samples are estuarine species for cytological, histological, and disease evaluation:

1. 100 to 200 oysters (*Crassostrea virginica*) from each major station by dredge or tongs (50 from each substation).

2. Otter trawl samples of flatfish, croakers, and other species will be taken from each major station. Probably, two to three trawls per major station will suffice. Fish caught in each trawl will be grossly examined both externally and internally by trained personnel. Fish will be enumerated and preserved for detailed histological examinations. All specimens will be brought to the EPA Gulf Breeze Laboratory for processing and examination.
3. Representative tissues from oysters and fishes will be processed for chemical analyses to determine concentrations of carcinogens.
4. Water samples will be taken monthly at each substation and analyzed by mass spectroscopy and liquid chromatography for chemicals that are suspect or actual carcinogens.
5. Sediment samples will be taken monthly from each substation and analyzed for carcinogenic chemicals.

Preceding and concurrent with the early part of the field sampling, a survey will be conducted to construct a profile of the chemical and pollutant entities which have been or are being released into waters of or adjacent to the major sampling station. The effort of this survey will be to obtain data from EPA regional offices, from state departments of environmental regulation, from municipal and county agencies, and finally from industry, when voluntarily appropriate and permissible. The profile produced will be used to decide what chemicals should be analyzed in this survey.

References

[1] Goldberg, Edward D.; Bowen, Vaughan T.; Farrington, John W.; Harvey, George; Martin, John H.; Parker, Patrick L.; Risebrough, Robert W.; Robertson, William; Schneider, Eric; and Gamble, Eric. The mussel watch. *Environmental Conservation* 5(2):101-125 (1978).

[2] Kidder, Gayle M. *Pollutant Levels in Bivalves: A Data Bibliography*. Scripps Institute of Oceanography, Mussel Watch Program. Washington: Environmental Protection Agency, 1977, EPA Contract R-80421501.

[3] Couch, John A.; Courtney, Lee A.; Winstead, James T.; and Foss, Steven S. "The american oyster (*Crassostrea virginica*) as an indicator of carcinogens in the aquatic environment." *Proceedings of the Symposium (1977) on Pathobiology of Environmental Pollutants: Animal Models and Wildlife as Monitors*. Washington: National Academy of Sciences, National Research Council, in press.

[4] Kidder, Gayle M., comp. *Pollutant Levels in Bivalves*—A Data Bibliography. Scripps Inst. Oceanogr., Mussel Watch Prog. EPA Cont. R-80421501.

8
Use of Benthic Macroinvertebrates as Indicators of Environmental Quality

D.R. Lenat, L.A. Smock, and D.L. Penrose

The benthic macroinvertebrate community frequently has been used to assess environmental quality. Benthic community structure can be used to indicate both the magnitude and the probable cause of environmental stress. A wide variety of procedures have been developed to evaluate benthic data, including the saprobic system, indicator organisms, indicator communities, reference station methods, diversity indices, and biotic indices.

Diversity indices, especially the Shannon-Weiner index, have been widely utilized. However, this index is insensitive to many pollutants, including inorganic particulates, pesticides, heavy metals, pH changes, and heat. Biotic indices may be used to evaluate environmental quality under a wider variety of conditions and are useful in summarizing large amounts of ecological data. Such indices, however, are valid only in specific geographic areas and should be supplemented by other analytical techniques. A biotic index for the North and South Carolina area is currently being developed by the North Carolina Department of Natural Resources and Community Development (NRCD) and includes information on greater than 1,000 species.

The symptoms of water "pollution" usually are caused by changes in the biotic components of the aquatic environment; yet, environmental stress is usually measured in terms of chemical parameters. The sole use of a chemical approach in describing water pollution not only is indirect but has also been found to be deficient.

The enforcement of water quality standards is generally the responsibility of state government systems. In order to evaluate water quality, state governments are becoming aware of the fact that analysis of the biological community is a valuable tool. The North Carolina Division of Environmental Management (DEM) has initiated many studies of the "biological integrity" of water bodies within the state; DEM biologists are responsible for collection and analysis of data related to water pollution problems. In several instances, a biological monitoring program has revealed problems undetected by using chemical-physical analyses.

The need for a biological approach has become generally accepted. The problem presently before us is the development of the theory, mechanics, and application of a biological approach to assessing water quality. To this end, the use of aquatic organisms, especially benthic macroinvertebrates, is presented as a useful technique.

Introduction

The reasons for using biological, rather than chemical, parameters to monitor water quality have been discussed frequently. Chemical studies do not integrate possible fluctuations in water quality between sampling periods; therefore, short-term critical events may often be missed. The biota, however, reflect both long- and short-term conditions. Since most species in a macroinvertebrate community have life cycles of a year or more, the effects of a short-term pollutant generally will not be overcome until the following generation appears. Macroinvertebrates are useful biological monitors because they are found in all aquatic habitats and are of a size which makes them easily collectible. Another important point is that chemical and physical analysis for a complex mixture of pollutants generally is not feasible. The aquatic biota, however, integrate and monitor the effects of a wide array of potential pollutants, including synergistic or antagonistic effects. For example, Patrick (1949) conducted a study of the Conestoga Basin using both chemical and biological sampling. Twenty-three streams were found to be "polluted" according to the biological data, but one-fourth of these streams had no signs of chemical pollution.

Analysis Systems

The first attempt at utilizing the benthos to provide information on water quality was the classic work of Kolkowitz and Marrson (1909) and their "Saprobien system." They developed lists of organisms which were associated with various zones of pollution, differentiated according to the degree of self-purification of organic matter in the system. These zones ranged from the polysaprobic (much decomposable organic matter and a low-dissolved oxygen concentration), through the alpha and beta mesosaprobic zones of recovery, to a clean-water oligosaprobic zone. These saprobic-system zones were considered as the "centers of optimum growth and development" for the organisms associated with them. An investigator could collect and identify the organisms present at a given location and through comparison with the established lists determine what degree of organic pollution was exhibited at that location.

This system, frequently refined and widely used in Europe (Thomas 1944; Kolkowitz 1950; Liebmann 1951; Sladecek 1965), has problems that limit its general utility. The system was based on the effects of organic matter decomposition in a smooth-flowing stream and focused on the tolerances of organisms to low-dissolved oxygen concentrations. It did not take into account the myriad forms of pollution present in today's rivers or the mediating effects of turbulence. Areas with increased water turbulence tend to have macroinvertebrate communities somewhat different from those associated with smoothly flowing streams. The major criticism, however, was that many species were not restricted to a single zone but could tolerate a wide range of environmental conditions. This was true for the organisms listed under the mesosaprobic zone. The mere presence or absence of an individual of a species provides little information on existing water quality conditions. The qualitative and quantitative changes in the entire community are of greater importance (Hynes 1960). Kolkowitz and Marrson (1909) note that "the main emphasis in the evaluation of waters should in general not be laid to the individual organisms but on the biocenoses...."

The major significance of the saprobic system was the introduction of the broad concept of indicator organisms. Indicator organisms are those organisms which, by their presence or absence, can provide information on the quality of a given body of water. Most investigators use the presence of certain species as indicators. For example, an aquatic community dominated by certain tubificids (especially *Tubifex tubifex* and *Limnodrilus hoffmeisteri*) or by midge larvae of the genus *Chironomus* can reflect an area with low-dissolved oxygen concentrations and high organic enrichment. Thienemann (1925) characterized the trophic states of lakes according to the midge larvae present: species of *Tanytarsus* were found in unproductive oligotrophic lakes and species of *Chironomus* were present in the oxygen-poor hypolimnetic waters of eutrophic lakes. However, as stated earlier, the presence of an indicator species does not in itself provide sufficient information on water quality. Organisms have a wide range of tolerance to pollution conditions (Hynes 1960). It can be deduced, therefore, that an observed absence of nontolerant species is of greater significance in indicating water quality than is the presence of tolerant species (Wurtz 1955).

It is at this point that a biologist's training is most useful. It is necessary to determine the exact species present, their normal life-history characteristics, and their responses to a broad range of environmental conditions. It is also helpful to examine physical characteristics of the body of water being investigated, especially in terms of substrate, flow, and temperature. The presence or absence of individuals or even of populations of "indicator species" is in itself insufficient for characterizing water quality. For example, a population of stoneflies, a group quite intolerant of organic wastes, generally indicates good water quality. However, the absence of stonefly populations does not necessarily indicate the reverse, since stoneflies may be limited to winter months and found only in riffle areas. In a classic study of the organically polluted Lytle Creek (Gaufin and Tarzwell 1956), none of the species common in the highly polluted area were restricted to that area. Likewise, occasional specimens of clean-water taxa could be found in oxygenated riffle areas of the recovery zone. Thus, rather than attempt to utilize the narrow and absolute idea of the presence or absence of indicator species, it is more informative to investigate "indicator communities."

The concept of indicator communities can be viewed from both qualitative and quantitative aspects. Qualitative methods are limited to determining what species are present, while quantitative methods determine the abundance and distribution of such species as well. The former method is a more strict definition of indicator communities and is essentially similar in concept to the idea described by Kolkowitz and Marrson (1908, 1909). Rather than noting the presence or absence of an indicator species, a determination is made of what species are associated together. For example, Milbrink (1973) discussed oligochaete communities as indicators of the trophic state in Swedish lakes. The classic indicator organism *Tubifex tubifex* was ubiquitously distributed, but associations of various oligochaete species had merit in indicating the trophic state.

Streams polluted by organic wastes have highly characteristic communities (figure 8-1). In "clean" waters, we find species richness characterized by numerous stoneflies, mayflies, caddisflies, and dipterans. Polluted areas usually have some species normally found in the cleaner areas but generally have an overall species composition different from the clean-water area. This species composition differs both qualitatively and quantitatively, suggesting an investigation of the structure of aquatic communities under natural and stressed conditions. The structure of natural communities has been found to follow the pattern of

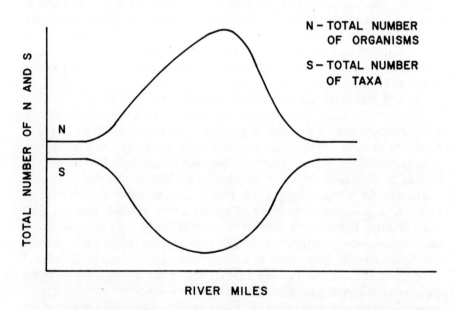

Figure 8-1. Typical Effects of Organic Pollution on Benthic Macroinvertebrates

a truncated curve (Patrick, Hohn, and Wallace 1954). This model suggests that there should be many species in clean, natural waters but represented by only a few individuals per given area. Also, there should be many rare species.

From this model, the concept of changing community diversity under different conditions of water quality was developed. The quantification of community structure, in the form of diversity and redundancy indices, was the next step in the utilization of aquatic organisms as indicators of water quality.

Before we proceed to a development of indices for community structure, it is important to note that we have been concerned primarily with instances of organic pollution. Other forms of pollution can have markedly different effects on various species within the community structure. The latter point is developed in greater detail below by including examples of species responses to nonorganic pollutants.

Environmental Stresses on Benthic Macroinvertebrates

Sediment

The input of inorganic particulate materials may affect benthic organisms from both habitat destruction and physical damage to the biota. Sediment generally results from nonpoint sources making it more difficult to evaluate cause and effect. Sediment decreased density to a greater degree than species richness (Chutter 1969; Cordone and Kelley 1960; Gammon 1970; Pearson and Jones 1975; Rosenberg and Snow 1975; Tebo 1955). It is possible, therefore, to differentiate sediment stress from the effects of toxic materials and organic wastes. Also, the dominant organisms are often characteristic of sediment pollution (Lenat, Penrose, and Eagleson 1979). The effect of sediment pollution varies according to stream flow (Lenat, Penrose, and Eagleson, in prep.). Under high-flow conditions, sand areas constitute a highly unstable habitat. Rubble areas may be considered "islands" of productive habitat in the midst of a "sea" of unproductive sand areas. As sediment inputs increase, the number and size of the islands decrease, accompanied by a marked decrease in density. Community structure, however, may show only minor changes. Under low-flow conditions the sand areas may form a stable substrate and develop distinctive fauna. Density in stable-sand areas may be greater than the density of control areas, but species richness is usually low.

Temperature

The impact of temperature elevation, especially as a result of power plant effluents, has been studied extensively. Moderate temperature increases rarely have demonstrable effects on the benthos (Langford and Aston 1972; Markowski 1960), but temperatures in excess of 30°C may result in substantial changes in the benthic fauna (Gallup, Hickman, and Rasmussen 1975; Lenat 1978). Data of Lenat (in prep.) suggest that thermal damage occurs only if maximum summer temperatures are well in excess of the "natural" maximum.

Toxic Materials

The reaction of the benthic communities to toxic materials differs markedly from the reaction to organic wastes. The difference has been illustrated by the ordination technique of Hocutt (1975). Severe doses of toxic materials will depress both density and species richness (figure 8-2). The remaining tolerant species often may be characteristic of a particular stress. Less severe doses of the toxic material may have variable effects on density. For example, a reduction in predator intensity may actually show an increase in density relative to control areas. This type of reaction has been seen for both acid mine drainage (Dill and Rogers 1974; Hall 1951; Roback and Richardson 1969; Simmons 1972) and pesticides (Holden 1972; Cape 1966; Muirhead-Thompson 1971; Penrose, in prep.).

Certain metals also elicit a reaction typical of toxic materials. Changes in the benthic community resulting from trace-metal inputs have not, however, been well documented. High concentrations of metals are generally required to cause an acute toxic response in macroinvertebrates (Warnick and Bell 1969),

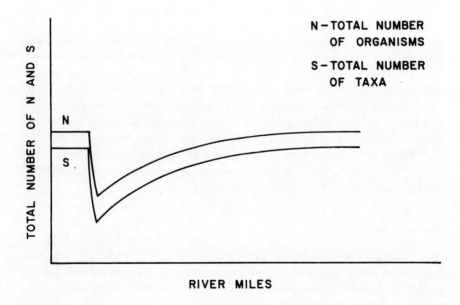

Figure 8-2. Typical Effects of Toxic Pollution on Benthic Macroinvertebrates

but chronic-toxicity studies have shown lethal and sublethal responses at very low levels (Spehar, Anderson, and Fiandt 1978). Wentzel, McIntosh, and Atchinson (1977) found a shift from chironomids to oligochaetes in a lake contaminated with cadmium, chromium, and zinc. Studies also have noted a decrease in numbers of all species and shifts in dominant species due to copper, lead, and zinc (Butcher 1955; Jones 1958; Winner et al. 1975). Whole-body concentrations in trace metals in macroinvertebrates tend to indicate concentrations in the surrounding water and sediment and therefore may be useful as indicators of long-term metal inputs (Schuman, Smock, and Haynie 1977). The usefulness of these organisms as indicators of short-term, pulse inputs of metals, however, is complicated by the rapid-loss rates of absorbed metals and because the major proportion of trace metals is associated with gut contents (Smock 1979).

Indices

Diversity Indices

Species diversity is widely utilized as an analytical tool, but the use of diversity indices has been criticized by many authors. Hurlbert (1971) labeled species diversity a "nonconcept" and questioned the use of diversity indices for both theoretical and practical reasons. Similar criticisms were made by Eberhardt (1976). Theory predicts that increasing diversity should lead to increased stability (Margalef 1968; MacArthur 1955), but this has not been borne out by field experiments (McNaughton 1968; Hurd et al. 1971).

Diversity indices have been used frequently in the analysis of benthic macroinvertebrate data, especially the Shannon-Weiner index $d = - \Sigma N_i/n_2 \log (N_i/N)$ and Margalef's (1951) index $d = S - 1/(\ln N)$ (with N = total density, N_i = density of a specific taxon, and S = species richness). Wilhm (1970) attempted to relate the Shannon-Weiner index to water quality and hypothesized that values greater than 3.0 indicate clean conditions, values less than 1.0 indicate severe pollution, and intermediate values indicate moderate pollution. However, this type of relationship appears to hold only for streams impacted by organic effluents: sewage, feedlot runoff, paper mill wastes, food processing wastes, and some oil refinery wastes. Organic pollution usually results in an increase in species richness and a decrease in density. Species richness decreases because of the elimination of sensitive species, especially those species intolerant of low-dissolved oxygen. Total density increases because the organic pollutant may be directly or indirectly used as food resource, and there may also be a reduction in predation and competition for the remaining species. These changes in both species richness and density contribute to a reduction in a calculated diversity index.

In more ecologically complex situations, diversity is not so closely linked to water quality (Godfrey 1978; Friberg et al. 1977). Many other pollutants may tend to decrease taxa richness and taxa abundance. Attempts to use a diversity index in these situations may give very confusing results. Gammon (1970) studied the effects of inorganic sediment on a small Indiana stream. He found that density was reduced up to 60 percent below the source of sediment, but the calculated diversity index did not change significantly. Similar patterns have been seen for a wide variety of other inorganic pollutants including insecticides (Odum 1969), metals (Winner et al. 1975), pH changes (Simmons 1972), and heat (Lenat 1978; Hocutt 1975).

Howmiller (1975) criticized the use of diversity indices because they reflect only the structure of the benthic community without respect to its composition. Thus, the index is not sensitive to the replacement of one species by another which may be ecologically very different.

Between-situation comparisons of diversity indices may also be complicated by "normal" environmental differences. For example, Olive and Dambach (1973) found that changes in diversity were related more to changes in substrate than to changes in water quality. Substrate has been shown by many investigators to influence species richness (Pennak and Van Gerpen 1947; Sprules 1947; Allan 1975). Greater substrate diversity, especially in riffle areas, usually is associated with greater species richness.

Several authors have recommended the use of species richness rather than species diversity as a measure of environmental quality (Dill and Rogers 1974; Winner et al. 1975). This approach, however, may also tend to yield misleading results. Species richness may decline in extremely oligotrophic conditions (Archibald 1972; Jumppanen 1976). Hilsenhoff (1977) also warns of "naturally low" species richness in some small cold streams. In this type of situation, it is preferable to examine changes in community composition, rather than changes in species richness.

In summary, the use of diversity indices is valid only where organic wastes are the only pollutants. Moreover, situations must be selected so that other differences, especially substrate changes, are minimized.

Biotic Indices

The *biotic index* is another analytical tool used to translate complex biological data into a simple measure of water quality. This type of index synthesizes known tolerance data for organisms with a quantitative measure of their abundance.

Several types of biotic indices have been developed. A simple index was devised by Beck (1953-1954). He assigned to each taxon a "quality factor" (from 1 to 5) according to its expected distribution along a gradient of environmental stress varying from clean water (0) to highly polluted (5). This index, however, utilizes only presence-absence data and neglects any quantitative information.

Woodiwiss (1964) used a different approach in his biological classification of the Trent River in England. Macroinvertebrate data were tabulated according to groups that vary from the species level to the family level. This number was combined with presence-absence data for six key indicator groups to derive (from a table) the index value. For example, the maximum value of ten is achieved if more than one species of Plecoptera is present (the most sensitive group), and more than sixteen total groups are present.

Chutter (1972) developed a biotic index (BI) for streams and rivers in South Africa. His index was the first to incorporate quantitative data and used the formula

$$\text{BI} = \sum_{S}^{i=1} \frac{n_i a_i}{N}$$

where n_i and a_i are, respectively, the density and quality factor of the ith taxon and N is the total number of taxa. Chutter's quality factor ranged from 0 (for species found in clean water) to 10 (for species in highly polluted water).

A similar index was developed by Hilsenhoff (1977) for the state of Wisconsin, but it uses quality factors ranging from 0 to 5. This index was based on a list of approximately 250 benthic macroinvertebrates. A biotic index is also being developed by Lenat and Penrose, North Carolina Division of Environmental Management (DEM), for the North and South Carolina area. The North Carolina index is currently based on a list of over 1,000 organisms and will eventually include data on 1,000 to 2,000 taxa.

The compilation of a list of organisms found (or expected to be found) is the first step in developing a biotic index. This immediately imposes a geographic limitation on use of the index. A single national or international listing would be not only too cumbersome but also inappropriate. An organism which is abundant in New England might be near the limit of its southern range in North Carolina and therefore under greater stress. Therefore, an index developed for the New England area might be expected to use a different quality factor for this organism than one developed for North Carolina.

The second step in developing a biotic index is to assign a quality factor to each taxon on the list. The quality factors assigned must summarize tolerance levels for a wide variety of pollutants, not just tolerance to organic wastes. This process is partly subjective, since it combines the experience of only one ecologist with supporting information available in literature. We feel that any subjective bias is diminished if the quality factors are derived by a group of ecologists rather than by one individual. To this end, the index developed by the authors for the North Carolina DEM was further verified and expanded at a miniconference held in May 1978. These meetings also serve to standardize taxonomy and compile a library of regional data.

The proper use of a biotic index requires the use of available taxonomy, preferably to a species level. As Resh and Unzicher (1975) have pointed out, different species within the same genus may differ sharply in their tolerance of environmental stress. In areas where species-level taxonomy is not possible, the biotic index concept is flexible enough to accept certain compromises.

It should be evident at this point that use of a biotic index requires considerably more preparation, at least initially, than use of a diversity index. The greater amount of information included in the biotic index, however, adds to its validity as a useful device for summarizing data.

The final step in development of the biotic index is to test it on actual data. Through this process, it is possible to associate index values with various levels of environmental stress. Hilsenhoff (1977) established five levels of stress varying from "clean, undisturbed" to "gross enrichment or disturbance." It should be possible at this stage to weight the index according to known geographic differences. For example, in the Carolinas, we might expect a "clean" stream in the coastal plain to have lower values than a "clean" stream in the mountain region. Field testing also enables the investigator to establish the number of samples required. This is easily done if a series of samples have been taken over approximately one year. The index is calculated for samples and added serially until the index is within a preselected percentage of the yearly mean. With enough data it also may be possible to weight the index by season. It is not expected, however, that a single sample will be adequate. Although the biotic index is intended to generate a quick and inexpensive assessment of environmental health, a single sample overlooks the seasonal variations resulting from flooding, input of pesticides, and so on. Since the benthic macroinvertebrates are being used to "monitor" the environment, the minimum sampling frequency should correspond to an average life span. Therefore, three to four samples per year should be adequate. Hilsenhoff calculated index values from samples of late spring, early summer, late summer, and late autumn and recommended certain "standard" correction values for samples from a single date.

A distinct advantage of the biotic index is that control data are essentially "built in." Establishment of control stations is often difficult because of natural environmental variation. By use of the biotic index, we can set up control data by making a rough estimation of the expected index value in an unstressed environment. This does not mean to suggest, however, that control stations should not be set up where feasible.

Biotic indices have been validated by several authors through comparisons of index values with chemical data. Hilsenhoff (1977), Chutter (1972), and Ghetti and Bonazzi (1977) found that changes in biotic index values were associated with a wide variety of chemical parameters. Furthermore, the biotic index was more closely associated with water quality than was a diversity index. A similar conclusion was reached by Mason (1975), using "hypothetical" chironomid data.

The concept of the biotic index has considerable utility in summarizing complex biological data for transmission to a nonprofessional audience. Thomas (1970) indicates a need for data transmission to groups that include regional planners, judiciary, legislatures, regulatory agencies, scientists, engineers, and special interest groups. Woodiwiss (1964) states that "often a long biological report is not fully comprehensible to other professions, inspectors and administrators, and the biologist is frequently asked for an understandable description in a few, well-chosen words. This is especially the case where the biological information is requested only as ancillary information."

We do *not* mean to suggest that a biotic index should be used by itself. It will be most useful when accompanied by standard chemical sampling and biological analysis. Supplementary biological analysis would focus on between-station differences by using such techniques as percentage similarities (Johnson and Brinkhurst 1971), cluster analysis (Crossman, Kaesler, and Cairns 1974), or ordination (Erman and Helm 1971; Hocutt 1975). Examination of changes in specific taxa or groups should be examined with regard to habitat preferences, feeding types, or special sensitivities. There is no substitute for an imaginative, well-informed professional analysis.

North Carolina Division of Environmental Management Analysis System

Biologists within the North Carolina DEM have developed an analysis system used to describe "biological integrity" within North Carolina (Penrose and Lenat 1978). The system is based on three levels of analyses. The first level describes data using single number summaries such as the Shannon-Weiner diversity index, the biotic index, percentage reduction in density from control data, and percentage reduction in species richness from control data. Level 1 analysis criteria are used to indicate levels of environmental stress. The first-level analysis has proved useful in presenting biological data to a nonbiological audience. The second level of analysis summarizes density and species-richness data for the major taxonomic groups (mostly orders of aquatic insects). These data, when compared to control data, can be utilized to assess changes on the aquatic community and to deduce the nature of pollutants. The biological monitoring group of DEM has also generated data from unstressed streams and rivers for many of the geographical regions in North Carolina. These series of control data may be substituted for control data if proper control sites cannot be located and to validate the use of control data. The third level of analysis summarizes data at the species level. Information on pollution tolerance, feeding type, emergence data, and so on of any given taxon, combined with qualitative data on its distribution, can often be related to specific chemical and physical changes in the environment.

Summary and Conclusions

Aquatic biologists have repeatedly demonstrated the utility of benthic macroinvertebrate data for water quality assessment. Chemical samples may appear to give greater precision, but the use of this approach may result in gross errors resulting from unmeasured pollutants, temporal variations, and synergistic effects. Each benthic species has a unique set of "niche" requirements; therefore, changes in environmental conditions will result in measurable changes in the benthic community. These changes may be interpreted both to determine the degree of stress and to deduce probable stress factors.

Many systems have been proposed over the last hundred years for the analysis of benthic data, including several indices. The Shannon-Weiner diversity index is currently the most widely used. This index, however, is insensitive to many types of pollutants. The development of regional biotic indices may be more useful in making environmental assessments. A "biotic" index of this type has been recently developed and tested by the North Carolina Department of Natural Resources and Community Development. This chapter describes this index and its utility in biological monitoring of pollutants.

References

Allan, J.D. 1975. The distributional ecology and diversity of benthic insects in Cement Creek, Colorado. *Ecology* 56:1040-1053.

Archibald, R.E.M. 1972. University in some South African diatom associations and its relationship to water quality. *Water Research* 6:1229-1238.

Beck, William M. 1953-1954. Studies in Stream Pollution Biology. I: A simplified ecological classification of organisms. *Quarterly Journal of the Florida Academy of Science* 17:211-227.

Butcher, R.W. 1955. Relation between the biology and the polluted condition of the Trent. *Verh. Internat. Verein. Limnog.* 12:823-827.

Chutter, F.M. 1969. The effects of silt and sand on the invertebrate fauna of streams and rivers. *Hydrobiologia* 34:57-76.

_____. 1972. An empirical biotic index of the quality of water in South African streams and rivers. *Water Research* 6:19-30.

Cordone, A.J., and Kelley, D.W. 1960. The influence of inorganic sediment on the aquatic life of streams. *California Fish and Game Journal* 47:180-228.

Crossman, J.W.; Kaesler, R.L.; and Cairns, J., Jr. 1974. The use of Duster analysis in the assessment of spills of hazardous materials. *American Midland Naturalist* 92:94-114.

Dill, G., and Rogers, D.T.; Jr. 1974. Macroinvertebrate community structure as an indicator of acid mine pollution. *Environmental Pollution* 6:239-262.

Eberhardt, L.L. 1976. Qualitative ecology and impact assessment. *Journal of Environmental Management* 4:27-70.
Erman, D.C., and Helm, W.T. 1971. Comparison of some species important values and ordination techniques used to analyze benthic invertebrate communities. *Oikos* 22:240-247.
Friberg, F.; Nilsson, L.M.; Otto, C.; Sjötōm, P.; Svensson, B.W.; Svenson, Bj.; and Ulfstrand, S. 1977. Diversity and environments of benthic invertebrate communities in south Swedish streams. *Archives of Hydrobiology* 81:192-254.
Gallup, D.N.; Hickman, M.; and Rasmussen, J. 1975. Effects of thermal effluents and macrophyte harvesting on the benthos of an Alberta lake. *Verh. Internat. Verein Limnog.* 19:552-561.
Gammon, J.R. 1970. The effect of inorganic sediment on stream biota. *Water Pollution Control Research Series*, EPA 18050DWC12170.
Gaufin, A.R., and Tarzwell, C.M. 1956. Aquatic macroinvertebrate communities as indicators of organic pollution in Lytle Creek. *Sewage and Industrial Wastes* 28:906-924.
Ghetti, P.F., and Bonazzi, G. 1977. A comparison between various criteria for the interpretation of biological data in the analysis of the quality of running waters. *Water Research* 11:819-831.
Godfrey, P.J. 1978. Diversity as a measure of benthic macroinvertebrate community response to water pollution. *Hydrobiologia* 57:111-122.
Hall, R.E. 1951. Comparative observations on the chironomid fauna of a chalk stream and a system of acid streams. *Journal of the Society of British Entomology* 3:253-262.
Hilsenhoff, W.L. 1977. "Use of Arthropods to Evaluate Water Quality of Streams." Technical Bulletin Number 100. Madison, Wisc.: Department of Natural Resources.
Hocutt, C.H. 1975. Assessment of a stressed macroinvertebrate community. *Water Resources Bulletin* 11:820-835.
Holden, A.V. 1972. The effect of pesticides on life in fresh waters. *Proceedings of the Royal Society of London B* 180:383-394.
Howmiller, R.P. 1975. "Analysis of Benthic Invertebrate Assemblages: Potential and Realized Significance for the Assessment of Environmental Impacts." In *Proceedings of the Nuclear Regulatory Commission Workshop*, NR-CONF-002, Biological Significance of Environmental Impact, eds. R.K. Sharma, J.O. Buffington, and J.T. McFadden, pp. 151-172.
Hurd, L.E.; Mellinger, M.J.; Wolf, L.L.; and McNaughton, S.J. 1971. Stability and diversity at three trophic levels in terrestrial successional ecosystems. *Science* 173:1134-1136.
Hurlbert, S.H. 1971. The nonconcept of species diversity: A critique and alternative parameters. *Ecology* 52:577-586.

Hynes, H.B.N. 1960. *The Biology of Polluted Waters.* Toronto: University of Toronto Press.

Johnson, M.G., and Brinkhurst, R.O. 1971. Associations and species diversity in benthic macroinvertebrates of Bay of Quinte and Lake Ontario. *Journal of the Fisheries Research Board Canada* 28:1683-1697.

Jones, J.R.E. 1958. A further study of the zinc-polluted river Ystwyth. *Journal Animal Ecology* 27:1-14.

Jumppanen, K. 1976. Effects of waste waters on lake ecosystem. *Annales Zoologici Fennie* 13:85-138.

Kolkowitz, R. 1950. Oekologie der Saprobien.Über die Beziehungen der Wasserorganismen zur Umwelt. *Shriftenreihe des Vereime für Wasserbdonen und Lufthygiene* 4:64.

_____, and Marrson, M. 1908. Ecology of plant saprobia. *Reports of the German Botanical Society* 26a:505-519.

_____, and Marrson, M. 1909. Ecology of animal saprobia. *International Review of Hydrobiology and Hydrogeography* 2:126-152.

Langford, T.E., and Aston, R.J. 1972. The ecology of some British rivers in relation to warm water discharges from power stations. *Proceedings of the Royal Society London Biological Science Series* 180:407-419.

Lenat, D.R. 1978. "Effects of Power Plant Operation on the Littoral Benthos of Belews Lake, North Carolina." *In Energy and Environmental Stress in Aquatic Ecosystems* eds. J.H. Thorp and J.M. Gibbons, pp. 580-596.

_____; Penrose, D.L.; and Eagleson, H.W. 1979. "Biological Evaluation of Non-point Source Pollutants in North Carolina Streams and Rivers." Raleigh: North Carolina Division of Environmental Management.

Liebmann, H. 1951. *Handbuch der Frischwasser und Abwasserbiologie*, Bd. I. Jena, Oldenburg Blg.

MacArthur, R.H. 1955. Fluctuations of animal populations, and a measure of community stability. *Ecology* 36:533-536.

Margalef, R. 1951. "Diversidad de Especies en las Comunidades Naturales." *Proceedings Institute Biological Applications* 9:5.

_____. 1968. *Perspectives in Ecological Theory.* Chicago: University of Chicago Press.

Markowski, S. 1960. Observations on the response of some benthonic organisms to power station cooling water. *Journal of Animal Ecology* 29:349-357.

Mason, W.T., Jr. 1975. "Chironomidae (Diptera) as Biological Indicators of Water Quality." *Organisms and Biological Communities as Indicators of Environmental Quality.* Columbus: Ohio State University.

McNaughton, S.J. 1968. Structure and function in California grasslands. *Ecology* 49:962-972.

Milbrink, G. 1973. Communities of Oligochaeta as indicators of the water quality in Lake Hjälmaren. *Zoon*, 1:77-88.

Muirhead-Thomson, R.C. 1971. *Pesticides in Freshwater Fauna.* N.Y.: Academic Press.

Odum, E.P. 1969. The strategy of ecosystem development. *Science* 164:262-270.

Olive, J.H., and C.A. Dambach. 1973. Benthic macroinvertebrates as indexes of water quality in Whetston Creek, Morrow County, Ohio [Scioto River Basin]. *Ohio Journal of Science* 73:129-149.

Patrick, R. 1949. A proposed biological measure of stream conditions based on 9 surveys of the Conestoga basin, Lancaster Country, Pennsylvania. *Proceedings of the Academy of Natural Science, Philosophy* 101:277-341.

_____; Hohn, M.H.; and Wallace, J.H. 1954. *Notulae Naturae*. Philadelphia: Academy of Natural Sciences. No. 259.

Pearson, R.G., and Jones, N.V. 1975. The effects of dredging operations on the benthic community of a chalk stream. *Biological Conservation* 8:273-278.

Pennak, R.W., and Van Gerpen, E.D. 1947. Bottom fauna production and physical nature of the substrate in a northern Colorado trout stream. *Ecology* 28(1):42-48.

Penrose, D.L., and Lenat, D.R. 1978. "Biological Studies of Several Western North Carolina Streams and Rivers to Determine the Impact of Non-point Sources of Water Pollution." Biological Series 101, N.C. Division of Environmental Management.

Resh, V.H. and Unzicher, J.D. 1975. Water quality monitoring and aquatic organisms: The importance of species identification. *Journal of Water Pollution Control Federation* 47:9-19.

Roback, S.S., and Richardson, J.W. 1969. The effects of acid mine drainage on aquatic insects. *Proceedings of the Academy of Natural Science Philosophy* 121:81-99.

Rosenberg, D.M., and Snow, N.B. 1975. "Ecological Studies of Aquatic Organisms in the Mackenzie and Porcupine River Drainage in Relation to Sedimentation." Environment Canada, Fisheries and Marine Service, Technical Report Number 547.

Schuman, M.S.; Smock, L.A.; and Haynie, C.L. 1977. "Metals in the Water, Sediment and Biota of the Haw and New Hope Rivers, North Carolina, UNC." Water Resources Research Institute Report Number 124.

Simmons, G.M., Jr. 1972. A preliminary report on the use of the sequential comparison index to evaluate acid mine drainage on the macrobenthos in a preimpoundment basin. *Transactions of the American Fisheries Society* 101:701-713.

Sladecek, V. 1965. The future of the saprobity system. *Hydrobiologia*, 25. The Hague: Dr. W. Junk Publishers.

Smock, L.A. 1979. "Analysis of Factors Influencing Whole-Body Metal Concentrations in Aquatic Insects." Ph.D. dissertation, University of North Carolina, Chapel Hill.

Spehar, R.L., Anderson, R.L.; and Fiandt, J.T. 1978. Toxicity and bioaccumulation of cadmium and lead in aquatic macroinvertebrates. *Environmental Pollution* 15:195-208.

Sprules, W.M. 1947. An ecological investigation of stream insects in Algonquin Park, Ontario. University of Toronto Studies, *Biology Series* 56:1-52.

Tebo, L.B., Jr. 1955. Effects of siltation resulting from improper legging on the bottom fauna of a small trout stream in the southern Appalachians. *Progressive Fish and Culturist* 17:64-70.

Thienemann, A. 1925. Die Binnengewässer Mitteleuropas. Eine Limnologische Einfuhrung. Binnengewässer, I-255.

Thomas, E.A. 1944. Versucheüber die Selbstreinigung fliessenden Wassers. *Mitteilongen aus dem Gebiete der Lebensmittelunteruchung und Hygiene* 35: 199-216.

_____. 1975. Indicators of environmental quality: An overview. *Indicators of Environmental Quality*, pp. 1-5. N.Y.: Plenum Publishing Co.

Warnick, S.L., and Bell, H.L. 1969. The acute toxicity of some heavy metals to different species of aquatic insects. *Journal of Water Pollution Control Federation* 41:280-284.

Wentzel, R.; McIntosh, A.; and Atchinson, G. 1977. Sublethal effects of heavy metal contaminated sediment on midge larvae (*Chironomus tentans*). *Hydrobiologia* (in press).

Wilhm, J.L. 1970. Range of diversity index in benthic macroinvertebrate populations. 42:R221.

Winner, R.W.; Van Dyke, J.C.; Caris, N.; and Farrel, M.P. 1975. Response of the macroinvertebrate fauna to a copper gradient in an experimentally polluted stream. *Verh. Internat. Verein. Limnoy.* 19:2121-2127.

Woodiwiss, F.S. 1964. The biological system of stream classification used by the Trent River Board. *Chemistry and Industry* 11:443-447.

Wurtz, C.B. 1955. Stream biota and stream pollution. *Sewage and Industrial Wastes* 27:1270-1278.

Research Suggestions—Benthic Invertebrates as Biological Indicators

John C. Morse

The use of benthic invertebrates ("benthos") for evaluation of the quality of surface waters has long been recognized as one of the most valuable tools for monitoring aquatic ecosystems. Current trends for evaluation of pollution levels in such ecosystems by the study of macroinvertebrates include examination for "indicator organisms" (Weber 1973), calculation of various diversity indices (Peet 1974), and calculation of "biotic indices" (Chutter 1972; Hilsenhoff 1977). Such techniques are superior to evaluations which rely solely on physical and chemical data because they measure the combined biological effects of all past, short-term pollutional stresses rather than just specific characteristics of the water at the time of water sampling.

The two most serious research needs in the use of macroinvertebrates for water quality monitoring are the necessity for the development of (1) standard, statistically defensible bottom-sampling technology and procedures and (2) species-level identification aids for use by monitoring scientists and technicians who may have only a minimum of training in benthos taxonomy.

The state of the art in benthos sampling is such that an inordinately large number of samples must be collected and analyzed to facilitate reducing the statistical standard deviation to levels that permit recognition of significant faunistic differences. This is especially true in lotic (flowing-water) habitats in which the substrate and flow characteristics tend to be highly variable from one sampling spot to another such that species, having different microhabitat preferences, tend to be highly clumped in their distribution.

Making adequate insect identifications is now complicated by the relatively recent discovery that species-level determinations are important and that more generic-level identifications are insufficient for assessing water quality and the impact of water pollution. For example, Gordon and Wallace (1975) demonstrated that although the caddisfly genera *Hydropsyche* and *Cheumatopsyche* are distributed through the length of the Savannah River Basin, many of the species in each genus are confined to portions of streams having only specific environmental characteristics. Temperature and dissolved oxygen were the most critical factors limiting the different distributions of the various species. Resh and Unzicker (1975) examined index genera of freshwater macroinvertebrates to which designations of "tolerant," "facultative," or "intolerant" had been assigned. Of eighty-nine such genera that have two or more species, sixty-one genera contain species that have tolerance levels other than that assigned to

113

each genus as a whole. Clearly, mere generic-level identifications are not adequate for meaningful water quality analyses.

Species-level identifications are difficult to accomplish because (1) existing comprehensive identification guides for the various taxonomic groups are often difficult to obtain and keep up to date, (2) the distribution of most aquatic species is poorly known, and (3) descriptions and keys for the majority of benthic immature stages of aquatic macroinvertebrate species simply do not exist (Wiggins 1966). Most species descriptions are based on adult forms of these animals which are either terrestrial or present in the water only for short periods of the year. Relatively little attention has been given to associating the adult forms with the benthic immature stages and describing and differentiating those aquatic forms.

Recommendation

Research funding priorities should be directed to research in quantitative benthic sampling technology and procedures and to basic taxonomic research with freshwater insects and other benthic macroinvertebrates.

References

Chutter, F.M., 1972. An empirical biota index of the quality of water in South African streams and rivers. *Water Research* 6:19-30.

Gordon, A.E., and Wallace, J.B. 1975. Distribution of the family Hydropsychidae (*Trichoptera*) in the Savannah River Basin of North Carolina, South Carolina, and Georgia. *Hydrobiologia* 46:405-423.

Hilsenhoff, W.L. 1977. Use of arthropods to evaluate water quality of streams. *Wisconsin Department of Natural Resources Technical Bulletin* 100:1-15.

Peet, R.K. 1974. The measurement of species diversity. *Annual Review Ecological Systems* 5:285-307.

Resh, V.H., and Unzicker, J.D. 1975. Water quality monitoring and aquatic organisms: The importance of species identification. *Journal of Water Pollution Control Federation* 47:9-19.

Weber, C.I., ed. 1973. "Biological Field and Laboratory Methods for Measuring the Quality of Surface Waters and Effluents." Cincinnati: *National Environmental Resources Center,* Environmental Protection Agency.

Wiggins, G.B. 1966. The critical problem of systematics in stream ecology. In *Symposium on Organism-substrate Relationships,* pp. 52-58. University of Pittsburgh, PA Laboratory of Ecology.

Part III
Terrestrial Plant and Soil Monitoring

Contributing authors in part III describe research activities on terrestrial plant and soil organisms useful as monitors of the effects from air and soil pollutants. These tests, measurements, and observations are conducted under controlled and uncontrolled ambient air and soil conditions.

Various types and species of organisms sensitive to air and soil pollutants are identified. A wide range of responses can be used as indices of pollutant exposure such as foliar injury, change in biochemical metabolism, growth, yield, and tissue accumulation of pollutants.

A description is given of research on a potentially useful bioindicators found in soils. This study is evaluating the effects of metal pollutants (mercury and cadmium) and of sulfur dioxide on rate of oxidation of the radionuclide, tritium, by this bacterium.

10 Vegetation—Biological Indicators or Monitors of Air Pollutants

H.C. Jones and *W.W. Heck*

Plants have proved valuable as biological indicators of air pollutants because (1) they show relatively high sensitivity to air pollutants compared to animals; (2) they show a large amount of genetic variability in sensitivity, both among and within species, which facilitates selection of varieties, cultivars, or clones of extreme sensitivity that can be used as indicators of exposure to low levels of pollutants; (3) they exhibit foliar symptoms of injury characteristic of exposure to a particular pollutant; and (4) they are immobile. Plant responses which can be used as indices of pollutant exposure include foliar injury; reductions in photosynthesis, growth, yield, and quality; changes in enzymes and metabolites; and accumulation of elements. However, in selecting and utilizing biological indicators, it is important to define the purposes of the monitoring: Is it to provide a system for the early recognition of potential pollutant problems or identification of areas exposed to pollutants; or is it to provide estimates of permanent adverse effects on human welfare, such as economic, aesthetic, or ecologic damages? The latter is much more difficult to quantify.

Historically, plants have served to indicate the presence of a pollution problem or to identify impacted areas. Effective use has been made of plants around sources of pollution, especially for SO_2 and HF. Plants have also been used to define the extent and bioeffects from chemical spills and other chemical problems. Selected plants are effective indicators of area and regional photochemical problems. These types of uses have permitted investigators to better define problem areas and to determine some sensitive response measures. Once these answers have been obtained, continued biological monitoring is no more productive than continued chemical monitoring. The responses which have been most commonly used to describe pollutant effects or presence are foliar injury and foliar accumulation (F, SO_2, and heavy metals). Their quantification has been fairly straightforward.

To be of continuing value, plants should be developed as monitors, not just indicators. As monitors, the plants' responses should include quantitative measures that are known to correlate well with specified responses of plants which are of economic or aesthetic importance. Therefore, responses such as reductions in growth, yield, quality, or photosynthesis or metabolic changes which might result from chronic exposure to low levels of pollutants, either individually or in mixtures, will have to be utilized as measures of pollutant effects. An ideal biological monitoring system would be one for which data for air pollution effects on growth, yield, and quality of major species of economic and ecological importance could be obtained under conditions representative of ambient conditions for a region. However, isolation of these kinds of air pollution effects from those caused by other factors in the ambient environment has proved both difficult and costly. Air-exclusion devices, which permit side-by-side comparisons of growth and yield of plants grown in ambient air with those grown in air which has been filtered to remove pollutants, provide a means for isolating and identifying air pollution effects under essentially ambient conditions while at the same time minimizing differences caused by variations in climatic and edaphic factors.

Presently, the primary emphasis in air quality monitoring is on chemical or physical measurements of the pollutants as they relate to compliance with ambient air quality standards. Biological monitoring could and should be used for documenting the effectiveness of control strategies in mitigating effects and in revision of criteria. Support for biological monitoring may be difficult to obtain. Support initially needs to go to the development of a useful monitoring system. Once a system is operational, it should have widespread use. At this time the possibility of developing such a system is uncertain, but it should effectively monitor biological responses and thus be of greater value than physical monitors. The installation of such a system and its maintenance would not be cheap, but it probably would be no more expensive than the current monitoring systems and would provide a true biological measure.

Attached is a bibliography of publications pertaining to plants as indicators and monitors of air pollution. This list shows the use of plants as indicators of pollution stress. Several discuss the possible use of plants as monitors.

The development of good biological monitoring systems would require research in the following areas:

1. The development and testing of experimental chambers and procedures for use in understanding the response of vegetation to air contaminants
2. The effects of pollutants on the growth and development of plants exposed to variable acute and/or chronic doses
3. The effects of environmental and biotic stresses on the response of plants to pollutants
4. Evaluation of plants as sinks for air pollutants (cleansing agents)
5. Identification of useful plant indicators for specific pollutants in breeding programs for the development of more resistant varieties
6. Determination of basic biological mechanisms of plant responses to pollutant stress

Bibliography

Benedict, H.M., and Breen, W.H. 1955. "The Use of Weeds as a Means of Evaluating Vegetation Damage Caused by Air Pollution." In *Proceedings of the Third National Air Pollution Conference* Los Angeles, Calif. pp. 177-190.

Berry, C.R., and Heggestad, H.E. 1965. "Air Pollution Detectives." In *Science for Better Living: USDA Yearbook of Agriculture.* Washington: USDA, pp. 142-146.

Bobrov, R.A. 1955. The leaf structure of *Poa annua* with observations on its smog sensitivity in Los Angeles County. *American Journal of Botany* 42:467-474.

Brandt, C.S., and Holzel, U. 1961. "Problems of the Recognition and Evaluation of the Effects of Gaseous Air Impurities on Vegetation." Department of Health, Education, and Welfare Technical Report Number A61-37. Translated from R. Guderian, H. van Haut, and H. Stratmann, *Z. Pflanzenkr. Pflanzenpathol. Pflanzenschutz* 67(5):257-264 (1960).

Brennan, E.G.; Leone, I.A.; and Daines, R.H. 1964. Atmospheric aldehydes related to petunia leaf damage. *Science* 143:818-820.

Darley, E.F.; Dugger, W.M., Jr.; Mudd, J.B.; Ordin, L.; Taylor, O.C.; and Stephens, E.R. 1963. Plant damage by pollution derived from automobiles. *Archives Environmental Health* 6:761-770.

Gilbert, O.L. 1964. "Lichens as Indicators of Air Pollution in the Tyne Valley." In *Ecology and the Industrial Society Symposium,* pp. 35-47. Held in Swansea, April 13-16, 1964. British Ecological Society Symposium Number 5.

Glater, Ruth A.; Solberg, R.A.; and Scott, F.M. 1962. A developmental study of the leaves of *Nicotiana glutinosa* as related to their smog sensitivity. *American Journal of Botany* 49:954-970.

Haagen-Smit, A.J.; Darley, E.F.; Zaitlin, M.; Hull, H.; and Noble, W. 1952. Investigations on injury to plants from air pollution in the Los Angeles area. *Plant Physiology* 27:18-34.

Heagel, A.S.; Body, D.E.; and Heck, W.W. 1973. An open-top field chamber to assess the impact of air pollution on plants. *Journal of Environmental Quality* 2:365-368.

———, and Heck, W.W. 1974. Predisposition of tobacco to oxidant air pollution injury by previous exposure to oxidants. *Environmental Pollution* 7:247-252.

Heck, W.W. 1966. The use of plants as indicators of air pollution. *International Journal of Air and Water Pollution* 10:99-111.

———. 1968. Factors influencing expressions of oxidant damage to plants. *Annual Review Phytopathology* 6:165-188.

———. 1972. "The Use of Plants as Sensitive Indicators of Photochemical Air Pollution." In *International Symposium on Identification and Measurement of Environmental Pollutants, Proceedings,* pp. 320-324. Ottawa: National Research Council of Canada.

———, and Brandt, C.S. 1977. "Effects on Vegetation: Native, Crops, Forests." In *Air Pollution,* 3d ed., ed. A.C. Stern. New York: Academic Press, pp. 157-229.

———; Dunning, J.A.; and Hindawi, I.J. 1965. Interactions of environmental factors on the sensitivity of plants to air pollution. *Journal of Air Pollution Control Association* 15:511-515.

_____; Fox, F.L.; Brandt, C.S.; and Dunning, J.A. 1969. "Tobacco, A Sensitive Monitor for Photochemical Air Pollution." Department of Health, Education, and Welfare, NAPCA Publication Number AP-55.

_____, and Heagle, A.S. 1970. Measurement of photochemical air pollution with a sensitive monitoring plant. *Journal of Air Pollution Control Association* 20:97-99.

_____; _____; and Cowling, E.B. 1978. "Air Pollution: Impact on Plants." *Proceedings of the Soil Conservation Society of America,* August 7-10, 1977. Publication Number 132, pp. 193-202.

_____; Mudd, J.B.; and Miller, P.R. 1977. "Plants and Microorganisms." In *Ozone and Other Photochemical Oxidants,* pp. 437-585. Washington: National Academy of Science.

_____; Taylor, O.C.; and Heggestad, H.E. 1973. Air pollution research needs: Herbaceous and ornamental plants and agriculturally generated pollutants. *Journal of Air Pollution Control Association* 23:257-266.

Heggestad, H.E.; Burleson, F.R., Middleton, J.T., and Darley, E.F. 1964. Leaf injury on tobacco varieties resulting from ozine, ozonated hexene-1, and ambient air of metropolitan areas. *International Journal of Air and Water Pollution* 8:1-10.

_____, and Darley, E.F. 1969. "Plants as Indicators of the Air Pollutants Ozone and PAN." In *Air Pollution.* Proceedings of the First European Congress Effect of Air Pollution on Plants and Animals, Wageningen, April 22-27, 1968, pp. 329-335.

_____, and Heck, W.W. 1971. Nature, extent and variation of plant response to air pollutants. *Advances in Agronomy* 23:111-145.

_____, and Menser, H.A. 1962. Leaf spot-sensitive tobacco strain Bel W_3, a biological indicator of the air pollutant ozone. *Phytopathology* 52:735.

_____, and Middleton, J.T. 1959. Ozone in high concentrations as cause of tobacco leaf injury. *Science* 129:208-210.

Hindawi, I.J.; Dunning, J.A.; and Brandt, C.S. 1965. Morphological and microscopical changes in tobacco, pinto bean and petunia leaves exposed to irradiated auto exhaust. *Phytopathology* 55:27-30.

Jacobson, J.W., and Hill, A.C., eds. 1970. "Recognition of Air Pollution Injury to Vegetation: A Pictorial Atlas." Air Pollution Control Association, Pittsburgh, PA.

_____, and Feder, W.A. 1974. "A Regional Network for Environmental Monitoring: Atmospheric Oxidant Concentration and Foliar Injury to Tobacco Indicator Plants in the Eastern U.S." Experiment Station Bulletin Number 604. Amherst: University of Massachusetts.

Lacasse, N.L. 1971. "Assessment of Air Pollution Damage to Vegetation in Pennsylvania." University Park: Center for Air Environment Studies, Pennsylvania State University. CAES Publication Number 209-71.

———; Weidensaul, T.C.; and Carroll, J.W. 1970. "Statewide Survey of Air Pollution Damage to Vegetation—1969." University Park: Center for Air Environment Studies, Pennsylvania State University. CAES Publication Number 148-70.

Larsen, R.I., and Heck, W.W. 1976. An air quality data analysis system for interrelating effects, standards, and needed source reductions—Part 3, Vegetation injury. *Journal of Air Pollution Control Association* 26:325-333.

LeBlanc, F., and Rao, D.N. 1966. Réaction de quelques lichens et mousses épiphytiques à l'anhydride sulfreux dans la région de Sudbury, Ontario. *Bryologist* 69:338-346.

MacDowall, F.D.H.; Mukammal, E.I.; and Cole, A.F.W. 1964. Direct correlation of air pollution ozone and tobacco weather fleck. *Canadian Journal of Plant Science* 44:410-417.

Middleton, J.T. 1961. Photochemical air pollution damage to plants. *Annual Review of Plant Physiology* 12:431-448.

———; Kendrick, J.B., Jr.; and Darley, E.F. 1955. "Airborne Oxidants as Plant-Damaging Agents." In *Proceedings of the Third National Air Pollution Symposium,* Pasadena, Calif., pp. 191-198.

———; ———; and Schwalm, H.W. 1950. Injury to herbaceous plants by smog or air pollution. *Plant Disease Report* 34:245-252.

———, and Paulus, A.O. 1956. The identification and distribution of air pollutants through plant response. *Archives Industrial Health* 14:526-532.

Noble, W.M., and Wright, L.A. 1958. A bio-assay approach to the study of air pollution. *Agronomy Journal* 50:551-553.

Rao, D.N., and LeBlanc, F. 1967. Influence of an iron-sintering plant on the epiphytic vegetation in Wawa, Ontario. *Bryologist* 70:141-157.

Syke, E. 1969. Lichens and air pollution, A study of cryptogamic epiphytes and environment in the Stockholm region. *Acta Phytogeographics Suecica* 52:1-123.

Taylor, O.C. 1969. Effects of oxidant air pollutants. *Journal of Occupational Medicine* 10:485-492.

11 Biological Monitoring Techniques for Assessing Exposure

G.B. Wiersma, R.C. Rogers, J.C. McFarlane, and D.V. Bradley, Jr.

Traditionally, monitoring systems have relied on detection of a residue in an environmental sample. This gives the amount of chemical present and sometimes the chemical form, but it tells very little about the potential impact of the chemical on the environment. We must realize that the important question is not whether a chemical residue is present in the environment, but what is the impact that man is having on his environment or what is the significance of chemical residue to man's health and well being. In order to answer these questions, our exposure monitoring systems of the future should consider techniques which will move us closer to a measure of the actual impact of pollutants rather than measuring chemical residues in various environmental compartments.

Exposure monitoring systems designed to answer the questions asked above must include the use of appropriate biological monitoring techniques which, if chosen properly, are a critical component of any exposure assessment monitoring system. Advantages of biological monitors are as follows:

1. In dynamic media such as air and water, they provide a natural integrating function.
2. They can concentrate some pollutant levels 10^3 to 10^6 over ambient.
3. They can serve as a direct measure of biological availability.
4. They can serve as an early-warning system that human populations are in danger of exposure or are being exposed.

Biological monitors can be useful as indicators, accumulators, or indices that measure dose. Detailed below are some specific examples. Some are currently under development at the Environmental Monitoring and Support Laboratory, Las Vegas, while others are examples of techniques which could be modified and developed for eventual incorporation into a responsive, exposure assessment monitoring program.

There is a considerable body of literature and experience in using animals and plants as accumulators of various toxic compounds. Organisms such as clams, ducks, starling, eagles, a variety of fish, and earthworms have been used as accumulators of particular compounds in the environment such as trace metals and pesticides.

It is not possible to review all pollutants for which biological accumulators are known. A specific example such as cadmium will help illustrate this point. Mollusks have been shown to be accumulators of cadmium in the marine environment. Phillips related cadmium concentration in the common mussel (*Mytilus edulis*) with total industrial daily input to Port Phillip Bay, Western Port Bay, Victoria, Australia [15, 16]. Valliela, Banus, and Teal successfully used mussels and clams to determine cadmium contamination from sludge applied to marshes [21]. The mussels and clams did not seem to accumulate lead and zinc.

Leatherland and Burton reported that *Anodonta* spp. mollusks from the polluted Thames estuary had twenty times the cadmium levels of similar species from the Solent, a relatively clean coastal area [14].

Considerable information is available on the ability of agricultural plants to concentrate and accumulate cadmium and thereby act as bioaccumulators. Several forest tree species have been tested [12], and forest litter can concentrate cadmium [3, 24]. One of the more unusual and successful attempts to use vegetation for an accumulation of airborne cadmium is the work by Goodman and Roberts [11]. They found that the moss *Hypnum* spp. was highly useful in concentrating airborne cadmium. They also had success using the grass *Festuca* spp.

Terrestrial systems can also be monitored by use of animals. Van Hook found that earthworms were capable of concentrating cadmium up to 22.5 times over the cadmium concentration in soil [23]. Gish and Christensen also found concentrations of cadmium in earthworms significantly correlated with cadmium levels in soil [10].

Finally, Coughtrey and Martin found that cadmium is readily available to invertebrates through a producer-herbivore system [7]. They sampled the terrestrial mollusk *Helix aspera* at different distances from a smelting complex in England. They found a significant correlation with cadmium content and body weight for this mollusk and felt that it would be a useful biological monitor, but only if individuals of similar age and size were used.

The Environmental Monitoring and Support Laboratory, Las Vegas, has been developing extensive reviews on the applicability of a variety of organisms as biological accumulators. Included in these reviews are fourteen toxic elements: antimony, arsenic, beryllium, boron, cadmium, chromium, cobalt, copper, lead, mercury, nickel, selenium, tin, and vanadium. A wide variety of organisms are considered in addition to various body segments. These segments are being researched to determine their suitability as sites for toxic-element accumulation. Included are mammalian hair, nails, claws and hoofs, and a variety of plants.

Biological accumulators are valuable for assessing retrospective exposures, as measures of accumulative exposure, indicators of critical transport pathways, and early-warning systems by accumulating toxic elements to levels detectable by current analytical techniques.

The use of accumulators is not the endpoint in the development of biological monitoring techniques. True exposure assessment techniques at the community, organism, and suborganism levels must be developed. Large bodies of information exist on effects of pollutants on the various hierarchical categories listed above, but few efforts have been made to adopt and incorporate this information into operational monitoring systems.

Plants can be employed as indicators of pollution. For example, a large body of knowledge has been developed on the effects of sulfur dioxide and fluoride on vegetation. Some types of plants that have been studied and used as field indicators of pollution are tobacco, Scots pines, lichens, and certain epiphytic mosses [20]. Most of this work has relied on visible damage and usually at levels high enough that the use of a plant system in this particular way is not practical as an early-warning system. An area for future research would be the development of rapid and cost-effective techniques for quantifying changes in productivity levels of plant systems which would be indicative of exposure to pollutants, for example, the development of remote sensing devices which can detect changes in spectra from stressed plants prior to visible damage occurring in the plant. In this latter area the Environmental Monitoring and Support Laboratory, Las Vegas, has been working for two years in cooperation with the U.S. Geological Survey on the potential application of a Fraunhofer Line Discriminator system for identifying early stress symptoms in plants before other signs of pollution damage can be ascertained (McFarlane's unpublished data).

The use of plants, soil, forest litter, small mammals, and domestic cattle is being studied under a variety of field monitoring systems for their suitability as indicators of pollutants in the environment.

Smith et al. have described how rumen-fistulated steers are used at the Nevada test site as biological integrators of present and potential biological buildup of transuranics in the environment [19]. Crockett and Wiersma described how soils and plants are being used to monitor the potential environmental contamination from geothermal energy development in the Imperial Valley of Southern California [8]. Brown, Wiersma, and Hester used common garden plants as potential indicators of damage from jet fuel at the Atlanta airport.[1] Wiersma, Brown, and Crockett used a variety of plant species and forest litter as part of the development of a pollutant monitoring system for background areas.[2]

In addition to the techniques mentioned above, other procedures show promise and should be investigated for their use as indicators of pollutant-induced impact and estimate of exposure.

Biochemical changes or reactions have the potential of being estimators of dose resulting from exposure to pollutants. Little work has been done in this area on applying these biochemical techniques as monitoring tools, but there is a large body of information on biochemical reactions which have the potential of being applied and/or modified for use in environmental monitoring systems.

Biochemical measures can be more cost-effective, in many instances, than other techniques for assessing pollutants in the environment.

While most of the work on potential biochemical indicators of pollutant exposure has been done with animals and microorganisms, some work has also been done with plants. For example, the biosynthesis of cellulose in isolated oat cell-wall sections was observed to be inhibited by exposure to peroxyacteyl nitrate (PAN) [20]. Although this technique is probably too cumbersome for field monitoring, it serves a possible example of what might be done in this area.

Biological impact monitoring of natural and agricultural communities involves measuring biological effects on multiple species. Some effort has been expended on developing methods, systems, and mathematical analyses to assess pollutant-induced changes in communities. These have involved evaluating the species of organisms present, their populations and changes in population, and their distribution. Much of this work has been done for aquatic habitats. In addition, some potential techniques for assessing pollutant impact on terrestrial communities have been described. Development of community-level indices of pollutant exposure should be undertaken for selected exposure monitoring requirements.

Microbial bioindicators, whose sensitivity to given concentrations of toxic materials is known, are excellent for determining the biological availability of toxic materials. At the Environmental Monitoring and Support Laboratory, Las Vegas, we have been investigating the use of *Alcaligenes paradoxus,* a hydrogen-oxidizing microorganism as a bioindicator for toxic materials.

It is known that soils are capable of supporting numerous hydrogen-oxidizing bacteria. Such organisms include certain of the pseudomonads, mycobacteria, actinomycetes, and two genera of uncertain taxonomic affiliation, *Paracoccus* and *Alcaligenes* [6]. These microorganisms are facultative chemoautotrophs which can derive their energy from the oxidation of hydrogen (H_2) [17, 18, 4, 13, 2]. Evidence would indicate the hydrogen/tritium (HT) gas can apparently be oxidized similarly.

A series of hydrogen enrichment cultures were initiated to isolate the organism in soil responsible for conversion of HT to HTO in soil.[3] These tests indicated that only the yellow bacteria and the streptomyces isolated from soil were capable of the oxidation of HT to HTO. Because the yellow-pigmented bacteria appeared to be responsible for the majority of the tritium oxidation, they were more fully investigated.

Davis et al. reported that there were some five species of gram-negative, rod-shaped, hydrogen-oxidizing bacteria which contained yellow pigment (carotenoids) [9]. These included *Pseudomonas flava* (Niklewski), Davis; *P. palleronii* Davis; *Alcaligenes paradoxus* Davis; and two other unnamed strains designated as 450 and 363. Table 11-1 lists some morphological, physiological, and nutritional characteristics of the unknown tritium-oxidizing, yellow bacteria and compares them with characteristics of the above-mentioned species.

**Table 11-1
A Comparison of Some Characteristics of Other Gram-Negative Rod-Shaped, Yellow-Pigmented Hydrogen Bacteria without Tritium-Oxidizing Isolate**

	Pseudomonas Flava	Strain 450	*Pseudomonas Palleronii*	Strain 363	*Alcaligenes Paradoxus*	Unknown Isolate
Yellow pigmentation	+	+	+	+	+	+
Oxidase reaction	+	+	+	+	+	+
Catalase reaction	+	+	+	+	+	+
Motility	+	+	+	−	+	+
Tolerance to 20 percent O_2 with H_2	−	+	±[a]	+	(+)[b]	+
Hydrolysis of starch	−	−	−	−	−	−
Organotrophic denitrification	−	−	−	−	−	−
Utilization of:						
H_2	+	+	+	+	+	+
Glucose	+	+	+	+	+	+
D-xylose	−	(+)	−	−	+	+
L-arabinose	+	+	−	−	+	+
D-mannose	+	+	−	−	+	+
D-galactose	+	+	−	−	+	+
Sucrose	+	+	−	−	−	−
Mannitol	+	+	−	−	+	+
para-hydroxybenzoate	−	−	+	−	+	+
L-leucine	−	+	+	+	+	+
L-tryptophan	−	−	+	−	±[c]	+
Histidine	−	+	−	−	±[d]	+

Source: Taken largely from D.H. Davis, R.Y. Stanier, M. Doudoroff, and M. Mandel, Taxonomic studies on some gram-negative polarly flagellated "hydrogen bacteria" and related species, *Archiv für Mikrobiolie* 70:1-3.

[a] One of two strains is positive, the other negative for this character.

[b] Parentheses indicate that mutation and selection may be required before the indicated reaction occurs.

[c] Ten of eleven strains are known to be positive for this character.

[d] Seven of eleven strains are known to be positive for this character.

The possession of "degenerately peritrichous" flagella, generally one or two, which usually originated from the subpolar region of the cell and extended several times the length of the cell, and the production of an unpleasant odor when grown autogrophically were additional characteristics common to both *A. paradoxus* and the isolate. Therefore, on the basis of these and the previous comparisons, it was decided to consider our isolate synonymous with *A. paradoxus*.

Tritium (and hydrogen) utilization by the *A. paradoxus* organism is rapid, indicating that there is a tritium/hydrogen sink in soils containing these types of microorganisms. Thus, it should be possible, by using methods outlined by

Bradley [5] and Rogers et al. (see note 3), with few modifications, to determine the oxidation rate in various soils. This technique also provides us with a simple yet definitive method to determine the effects of parameters such as acidity, organic carbon sources, temperature, partial pressures of gases, and others on the oxidation reaction. The ability to measure this reaction rate may also eventually provide us with the means for using *A. paradoxus* as a bioindicator of toxic levels of compounds.

A study was conducted to determine the effect of mercury on the rate of oxidation of tritium (hydrogen) by the soil isolate of *A. paradoxus*. The hope was to use the above technique to rapidly analyze samples for the bioavailable concentration of the particular metal of interest, as opposed to current methodology which can only analyze for levels chemically available.

The effects of mercury were tested in both solutions and sterile soils. The solutions were prepared as follows.

Round-bottomed, 1-liter flasks were prepared with 0.025 M potassium phosphate buffer at pH 5.0, 7.0, or 7.2. These contents were then amended with mercuric nitrate [$Hg(NO_3)_2$] to yield the desired final concentrations. Controls received a volume of distilled water equal to the volume of the metal solution addition. Cells were then added to the flasks which were stoppered, injected, and shaken. The final volume of liquid within each flask was 15.0 ml. Collection of the water by benezene distillation and the analysis for its tritium content were done as described by Bradley [5].

Sterile soils were prepared as follows:

Twenty grams (20g) of steam-sterilized clay loam were put into each 1-liter, round-bottomed flask. Sterilization was verified by plating soil dilutions on Triptase Soy Agar (TSA). After this, 14.0 ml of the *A. paradoxus* cell suspension and 1.0 ml of the proper dilution of $Hg(NO_3)_2$ were added to yield a final concentration of 1.0, 10.0, or 100.0 parts per million (ppm) mercury. Controls received 1.0 ml of distilled water instead of the mercury solution. The flasks were then stoppered and injected, and the mixture was well shaken to create a "slurry" covering the entire inner surface of the flask. The flasks were then incubated without shaking in a room at 30°C. Samples were removed and the water distilled and analyzed as before.

The effect of a variety of mercury concentrations in solution, pH 7.2, on the ability of *A. paradoxus* to oxidize tritium is shown in figure 11-1. It is evident that *A. paradoxus* is very sensitive to somewhere between 0.1 and 1.0 ppm mercury in an uncomplexed solution. In the sterilized soil (figure 11-2) amended with *A. paradoxus* and mercury, 1.0 ppm was sufficient to completely inhibit the oxidation reaction but only after approximately 20 percent conversion of the tritium to tritiated water had taken place. For the same concentration of mercury in solution, only about 5 percent conversion took place. It is noteworthy that 100.0 ppm mercury in the soil produced the same effect as 1.0 ppm in solution. This suggests that the clay-loam soil offered some degree of protection not available in solution. This is consistent with the results obtained by Babich and Stotzky who found that cadmium toxicity was reduced

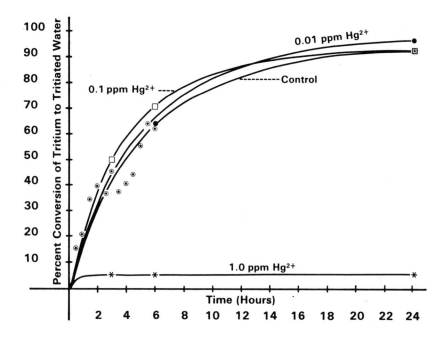

Figure 11-1. Tritium Oxidation by *Alcaligenes Paradoxus* in 0.025 M Potassium Phosphate Solution, pH 7.2

by clay minerals [1]. They determined that the degree of protection was directly related to the cation-exchange capacity of the clays used. Work by Van Faassen seems to indicate that this is true for mercury also since he found that in sandy soils the toxicity of mercury was greater than in clays soils [22]. Thus, since the soil used here was a clay loam, it should not be surprising that some protection from mercury was afforded to the bacteria. The negatively charged clay particles, which could be binding the Hg^{2+} ions with other cations in the solution, probably accounted for the eventual toxic effect on the bacteria.

A concern in determining the usefulness of *A. paradoxus* for future biological monitoring was that it might be resistant to low levels of heavy metals. For mercury this appears not to be the case.

Although this study is far from conclusive, it clearly shows that *A. paradoxus* is sensitive to low levels of mercury in both solution and soil. It seems that *A. paradoxus* has the potential, given sufficient research emphasis, of serving as a sensitive bioindicator of environmental mercury.

Other trace elements such as cadmium and lead have been tested and described by Bradley [5]. Similar results were obtained; however, pretreatment of the cells was needed to increase sensitivity of *A. paradoxus* to cadmium and lead.

The effect of sulfur dioxide (SO_2) on the hydrogen oxidation potential of *A. paradoxus* was tested under field conditions of the Colstrip Power Plant.

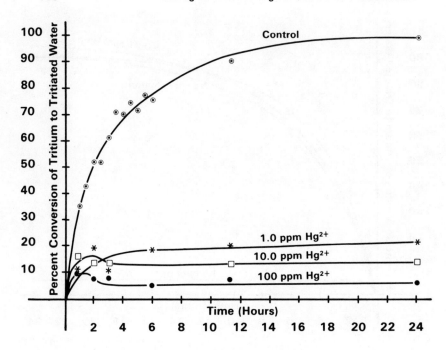

Figure 11-2. Tritium Oxidation by *Alcaligenes Paradoxus* in Sterilized Clay-Loam Soil

Field plots have been established and exposed to SO_2 fumigations of known concentration and duration. This is called the Zonal Air Pollution Study (ZAPS).

Soils from each of the ZAPS fumigation plots were anlayzed for their hydrogen oxidation potential by using *A. paradoxus* as a representative microorganism. In the ZAPS I soils, there was a significant decrease in activity present except that the microorganism activity in the highest SO_2 fumigation was significantly lower than all other plots. The results of these tests indicate that after one year of fumigation, only the most severe insult resulted in a detectable depression. However, after two years the soil microbiota was significantly affected by the SO_2 fumigation. These tests also indicate that the insult measurement technique employed might serve as a monitoring tool with sufficient sensitivity to detect this insult.

In summary, biological monitors of pollutant accumulation, transformations, and pathways need to be identified and developed for use in exposure monitoring. There is a specific need for rapid biological monitoring methods which are pollutant-specific or pollutant-class-specific and give a direct assessment of the impact of a pollutant on man and his environment.

Notes

1. K.W. Brown, G.B. Wiersma, and N. Hester, "Investigation of Soils and Vegetative Damage: Vicinity of Hartsfield, Atlanta's International Airport—Mountain View, Georgia." U.S. Environmental Protection Agency, Las Vegas, Nev., February 1978.
2. G.B. Wiersma, K.W. Brown, and A.B. Crockett, Development of a pollutant monitoring system for biosphere reserves and results of the Great Smoky Mountains pilot study. *Proceedings 4th Joint Conference on Sensing of Environmental Pollutants,* New Orleans, La., pp. 451–456.
3. R.D. Rogers, D.V. Bradley, and J.C. McFarlane, The role of a hydrogen-oxidizing microorganism, *Alcaligenes paradoxus,* in environmental tritium oxidation. In review.

References

[1] Babich, H., and Stotzky, G. Reductions in the toxicity of cadmium to microorganisms by clay minerals. *Applied and Environmental Microbiology* 33:696–705, 1977.
[2] Bernard, U.; Probst, I.; and Schlegel, H.S. The cytochromes of some hydrogen bacteria. *Archiv für Mikrobiolie* 95:29–37, 1974.
[3] Bolter, E.; Butz, T.; and Arsenau, J.F. "Mobilization of Heavy Metals by Organic Acids in the Soils of a Lead Mining and Smelting District." Presented at 9th Annual Conference on Trace Substances in Environmental Health, Columbia, Mo., June 9-12, 1975.
[4] Bongers, L. Phosphorylation in hydrogen bacteria. *Journal of Bacteriology* 93:1615–1623, 1967.
[5] Bradley, D.V. "Physiological and Morphological Studies on a Tritium ($^3H_2$)–Oxidizing Soil Microorganisms, *Alcaligenes paradoxus,* and Its Possible Use as a Biological Monitor." Master's thesis, Department of Biology, University of Nevada, Las Vegas.
[6] Buchanan, R.C., and Gibbons, N.E., eds. *Bergey's Manual of Derminative Bacteriology,* 8th ed. Baltimore, Md.: The Williams and Wilkins Co., 1974.
[7] Coughtrey, P.J., and Martin, M.H. The uptake of lead, zinc, cadmium and copper by the pulmonate mollusc, *Helix aspera* Muller, and its relevance to the monitoring of heavy metal contamination of the environment. *Oecologia* 27:65–74, 1977.
[8] Crockett, A.B., and Wiersma, G.B. Status of baseline sampling for elements in soil and vegetation at four KGRA's in the Imperial Valley, California. Geothermal Resources Council, *Transactions:* 65-67, 1977.

[9] Davis, D.H.; Stanier, R.Y.; Doudoroff, M.; and Mandel, M. Taxonomic studies on some gram-negative polarly flagellated "hydrogen bacteria" and related species. *Archiv für Mikrobiolie* 70:1-3, 1970.

[10] Gish, C.D., and Christensen, R.E. Cadmium, nickel, lead and zinc in earthworms in roadside soil. *Environmental Science Technology* 7(11):1060-1062, 173.

[11] Goodman, G.T., and Roberts, T.M. Plants and soils as indicators of metals in the air. *Nature* 231:287-292, 1971.

[12] Hemphill, D.D., and Pierce, J.O. "Accumulation of Lead and Other Heavy Metals by Vegetation in the Vicinity of Lead Smelters and Mines and Mills in Southeastern Missouri." In *Trace Contaminants in the Environment.* Proceedings of the Second Annual NSF-RANN Trace Contaminants Conference, Asilomar, Pacific Grove, Calif., August 29-31, 1974, pp. 325-332.

[13] Ishaque, M.; Donawa, A.; and Aleem, M.I.H. Energy-coupling mechanisms under aerobic and anaerobic conditions in autographically grown *Pseudomonas saccharophilia. Archives Biochemistry and Biophysics.* 159:570-579, 1973.

[14] Leatherland, T.M., and Burton, J.D. The occurrence of some trace metals in coastal organisms with particular reference to the Solent region. *Journal of Marine Biology Association of the United Kingdom* 54(2):457-468, 1974.

[15] Phillips, D.J.H. The common mussel *Mytilus edulis* as an indicator of pollution by zinc, cadmium, lead, and copper. I: Effects of environmental variables on uptake of metals. *Marine Biology* 38:59-69, 1976.

[16] Phillips, D.J.H. The common mussel *Mytilus edulis* as an indicator of pollution by zinc, cadmium, lead, and copper. II: Relationships of metals in the mussel to those discharged by industry. *Marine Biology* 38:71-80, 1976.

[17] Repaske, R. Characteristics of hydrogen bacteria. *Biotechnological Bioengineering* 8:217-235, 1966.

[18] Schlegel, H.D. "Physiology and Biochemistry of Knallgas Bacteria." In *Advances in Comparative Physiology and Biochemistry,* vol. 2, ed. O. Towenstein. New York: Academic Press, pp. 185-237.

[19] Smith, D.D.; Black, S.C.; Giles, K.R.; Bernhardt, D.E.; and Kinnison, R.R. *Tissue Burdens of Selected Radionuclides in Beef Cattle on and around the Nevada Test Site.* Environmental Monitoring and Support Laboratory, Las Vegas, Nev. Environmental Protection Agency, NERC-LV-539-29.

[20] Thomas, W.A., ed. *Indicators of Environmental Quality.* New York: Plenum Press, 1972.

[21] Valliela, I.; Banus, M.D.; and Teal, J.M. Response of salt marsh biovalves to enrichment with metal-containing sewage and retention of lead, zinc, and cadmium by marsh sediments. *Environmental Pollution* 7:149-157, 1974.

[22] Van Faassen, H.G. Effects of mercury compounds on soil microbes. *Plant Soil* 38:485-487, 1973.

[23] Van Hook, R.I. Cadmium, lead, and zinc distribution between earthworms and soils: Potential for biological accumulation. *Bulletin of Environmental Contamination Toxicology* 12(4):509-512, 1974.

[24] Van Hook, R.I.; Harris, E.F.; Henderson, G.S.; and Reichie, D.E. "Trace Element Distributions on Walker Branch Watershed." In *Trace Contaminants in the Environment*. Proceedings of the Second Annual NSF-RANN Trace Contaminants Conference, Asilomar, Pacific Grove, Calif., August 29-31, 1974, pp. 165-171.

Part IV
Ecological Monitoring

A keynote chapter by A. Hirsch defined the scope of ecological monitoring to include efforts for improving our understanding of "... the extent to which man's activities are altering the structure and function of natural ecosystems...."

An inventory conducted by the Oak Ridge National Laboratories identified an extensive national effort for obtaining baseline knowledge on ecological monitoring such as the time trends in populations of naturally occurring flora and fauna.

A description is given of a conceptual network of Environmental Research Parks that would serve as field laboratories in continually assessing the ecological impact of human activities.

12 Monitoring Cause and Effects—Ecosystem Changes

Allan Hirsch

This chapter begins by identifying those aspects of biological monitoring that are *not* discussed here. This discussion does *not* focus on biological monitoring done for the purpose of:

1. Using biological indices as a surrogate for chemical or other measurements to assess the degree of contamination (such as the use of benthic organisms to determine the shape of an oxygen sag downstream from an effluent discharge)
2. Measuring isolated biological impacts (for example, polychlorinated biphenyls (PCB) levels in fish, shellfish contamination by heavy metals, or the number of fish kills in the vicinity of effluent discharges)
3. Monitoring biological parameters as an index of human health—the "canary in the mine" analogy.

All these types of biological monitoring have an important role in pollution regulation and research. Further, many of them can contribute directly to or become components of ecosystem monitoring.

Instead, the subject of this discussion is somewhat different—and somewhat more elusive—monitoring to understand the status and trends of natural ecosystems! This topic deals with the monitoring necessary to determine the extent to which human activity is altering the structure and function of natural ecosystems such as rivers, lakes, and streams; forests, deserts, and prairie lands; and estuarine, coastal, and oceanic areas. In the broadest sense, it is monitoring to understand the human impact on the biosphere itself.

This distinction can be illustrated by discussing the 1972 amendments to the Federal Water Pollution Control Act (FWPCA). The amendments state that an objective of the national pollution control is "to restore and maintain the natural physical, chemical, and biological integrity of the Nation's waters." And "integrity" has been described as maintaining the structure and function of aquatic ecosystems. Although water quality monitoring has received much attention, as yet relatively little seems to have been accomplished to give the ecological objective real operational meaning by defining monitoring programs which would measure progress toward restoring and maintaining ecosystem integrity. Much of the current monitoring effort seems to focus on the levels and distribution of contaminants, rather than on the end ecological result.

There are several reasons why it is important to monitor the structure and function of natural ecosystems. Most important, the condition of these ecosystems may ultimately be the underpinning of the life support systems on which we all depend. Here, reference is made to their role in the earth's biogeochemistry. We need to be able to determine whether humans are altering carbon dioxide, oxygen balance, nitrogen cycling, and similar phenomena.

Or, to describe this need in less cosmic terms, these ecosystems provide important and often essential goods and services. Examples are fisheries production, grazing capacity of rangelands, watershed functions, timber production, and other benefits.

Last, in our affluent society, maintenance of natural ecosystems is very important to the quality of life. This is reflected in the environmental movement and is one of the underlying aims of the National Environmental Policy Act (NEPA). The Endangered Species Act is another example of national concern with ecosystem protection. Most recently, the President has established a National Heritage Conservation Program; one of its aims is to maintain diversity of natural ecosystem types as part of our national heritage.

In summary, then, ecosystem monitoring is an important aspect of environmental assessments from the standpoint of life support functions, goods and services, and quality of life.

Monitoring to assess ecosystem status and trends falls into several broad categories. First, we can measure the rate of destruction of certain ecosystem types. Obviously, pollution is not the only stress that impacts natural ecosystems. Many human-induced changes are basically physical in nature. Some examples are the destruction of natural water courses through water diversion, channel alteration, and other engineering developments; drainage or filling of wetlands; surface disturbance through mining, transportation, and other land-use changes; and deforestation through timbering or grazing.

If we are interested in status and trends of natural ecosystems, and not merely with the extent, distribution, and impacts of contaminants, we have to be concerned with these alterations. So one kind of ecological monitoring is basically an inventory effort to determine the quantity, distribution, and rates of loss (or gain) of various ecosystem types. Inventories designed to measure such major environmental problems as loss of tropical rain forests, increased desertification, or wetlands loss are examples.

This kind of status monitoring is fairly straightforward; it relies heavily on applications of remote sensing, through which it has become feasible to monitor changes taking place over very large areas. There is quite a lot of activity of this type, both nationally and internationally, and it is yielding valuable information about status and trends.

A second class of ecological monitoring involves monitoring to assess the extent and significance of specific known contaminants or stresses, usually on a site-specific geographic scale. Examples are the effects of power plants on the

Hudson River estuary, the effects of oil shale development in the Piceance Basin, Colorado, or the effects of sewage discharges on the Potomac River.

There are a myriad of such studies, conducted either in response to some specific regulatory or management requirement or for research purposes. For some problems, such as the impacts of organic pollution on stream biota, techniques of study and monitoring are well developed. However, when we consider larger, more open ecosystems, even these site-specific, problem-specific studies become quite difficult to design and interpret. One can see something of these difficulties in reviewing various studies of ocean waste disposal, such as the Southern California Water Quality Research Project or studies of ocean dumping in the New York Bight where, despite costly and intensive efforts, it has proved quite difficult to get conclusive answers of any sort. In general, as we move toward looking at larger, more complex systems and toward more subtle or less devastating effects, our understanding of how to monitor and of what is actually occurring in the ecosystem becomes much weaker. A few of the scientific and technical problems involved are discussed below.

The problem becomes most difficult when we look at a third category of ecological monitoring (and here it should be stressed that although we are talking of categories, in actuality the various monitoring situations represent a continuum). The third category is long-term monitoring to detect subtle changes or shifts in ecosystems, usually on a regional or widespread areal scale. An example is current concern about widespread effects of acid rain or other atmospheric deposition on forest ecosystems of northeastern United States and northern and central Europe. The impacts of widespread increases in low-level hydrocarbon contamination of the oceans on marine food webs presents another case in point. Here is where the early-warning concept becomes so important. We can measure the contamination, but can we see incipient changes in ecosystem response to these stresses? Can we detect these and take early action before the problem becomes catastrophic?

Finally, the problem of monitoring ecosystem status is most difficult when we have not even isolated a specific stress to worry about, when we merely want to monitor to know whether things are satisfactory ecologically or whether something is beginning to go wrong.

Let us briefly examine some of the difficulties in mounting monitoring programs to address this very important question: Are our ecosystems "healthy," or are we altering them, however subtly? There are obviously very real scientific and technical difficulties involved in monitoring ecosystem changes. There are some basic reasons for this.

First is the complexity and dynamism of ecosystems. Ecosystems are constantly in a state of flux, and they exhibit wide variability over space and time. This makes it very difficult to detect signals until those signals become very obvious. The warning signs may be masked by the background noise of natural variability.

Further, some of the natural stresses on ecosystems are perhaps of greater magnitude than human-induced stress. The impacts of Hurricane Agnes on Chesapeake Bay in terms of mortality to oysters and other marine life were very severe and may well have been greater than the cumulative impacts of pollution. When these natural stresses interact with human-induced stress, such as the interaction of drought and overgrazing in the Sahel, the results can be catastrophic. These natural stresses make it difficult to detect and separate certain human-induced stress, at least on an early-warning basis.

The Fish and Wildlife Service has been trying to develop a system for monitoring the ecological impact of oil shale development in the Piceance Basin of Colorado. One of the prime values in that ecosystem is the mule deer herd. The herd crashed a few years ago. This is now believed by many to be a result of several successive hard winters, with the loss of faunal crops. But it has been observed (facetiously, one should hasten to add) that if the crash only had happened subsequent to the more recent initiation of pilot oil shale development, we could blame it all on oil shale. So even when we do detect significant ecosystem changes, we often have a hard time linking them to the causal mechanism.

Then, even if we can show that a change is resulting from human activity, we have to ask ourselves how significant it is. Is it important? Irreversible? Catastrophic? We know that there are many compensatory mechanisms in ecosystems and that ecosystems exhibit varying degrees of resiliency. So even where we do detect that impacts are occurring, we sometimes have a hard time defining their real importance. Are they transient effects or unidirectional changes?

All this scientific complexity and uncertainty poses some basic problems in mounting research and monitoring programs to detect ecosystem change. Obviously, there are various approaches to scientific design that can help narrow down, though not eliminate, some of these difficulties. Time does not permit discussion of those approaches here.

However, these very real scientific difficulties are compounded by a number of institutional problems. The first of these involves funding mechanisms for long-term work. If we are to better distinguish from among long-term, widescale ecosystem changes—those which are cyclic, those which are unidirectional, those which are natural, and those which are human-induced—then we must maintain some basic, long-term ecosystem studies.

As yet, virtually no funding mechanisms are available to support a program of continuing, long-term ecological research and monitoring. There are current efforts to get such programs underway. Among these are the Man and the Biosphere (MAB) program, an international effort coordinated by UNESCO in which eight-seven countries are participating in studies and information exchange designed to address regional and global environmental programs. As part of the MAB program, a coordinated international network of protected areas, termed *biosphere reserves*, has been under development for the purpose of conserving genetic diversity and for use in ecological research and monitoring. At present 144 areas have been established, representative of the world's biological regions; 29 of these are in the United States. As yet, the amount of coordinated, long-term ecological monitoring initiated has been extremely limited. However, efforts to fund and implement a more vigorous effort are continuing.

In addition, the National Science Foundation has proposed a similar network, entitled Experimental Ecological Reserves, to include a wide range of representative ecosystems in which manipulative studies would be encouraged. However, no formal funding or organizational structure has been developed yet to ensure implementation, Similarly, the United Nation's Environmental Program's proposed Global Environmental Monitoring System, which would incorporate ecosystem monitoring of the type being discussed here, is not yet underway on a significant scale.

Some smaller-scale efforts at long-term ecological research and monitoring are being made. For example, the Department of Energy has four National Environmental Research Parks which you will be hearing more about in this book. These areas have been intensively studied, but they represent a very small sampling of the range of ecosystems across the United States. National Oceanic and Atmospheric Administration (NOAA) under the Coastal Zone Management Act, (NOAA) has been supporting establishment of a selected number of estuarine sanctuaries, to be managed by the states for the purposes of ecological research; this program is still in its early stages. Several other similar efforts could be cited. However, in the final analysis, we cannot yet say that we have anything even remotely approaching a long-term ecological monitoring network which would enable us to systematically detect incipient ecosystem trends.

Second, there often has been inadequate dialogue between the government agencies funding ecological studies, in response to NEPA or other regulatory or mission-oriented need, and the scientists doing the work. If anything, the traditional dialogue between the administrators and the ecologists has been a polarized one. The agency administrator concerned with resource management or environmental regulation wants *some* answer, the best answer he can get, by a certain date. The research ecologist has traditionally emphasized the difficulties of providing an answer, the unknowns, and above all the need for scientific freedom unconstrained by deadlines or output-oriented objectives. In fact, it sometimes seems that ecologists have an occupational disease of wanting to stress the unknowns, rather than the knowns. There is a desire to stress that every situation in nature is unique or individual, and therefore, they feel a reluctance to generalize.

Yet, we should recognize that decision making is done in an atmosphere of uncertainties—economic and technological as well as biological. If ecological considerations are to contribute more efficiently to the decision process, then we must be willing to help improve the probabilities, even if we are not sure.

At any rate, in recent years the many millions of dollars spent in response to NEPA and various needs have provided a unique opportunity to improve our understanding of ecosystem change. Instead, many of these funds have been spent on safe, descriptive data collection—studies which are often misguided, misdirected, and mediocre. Such studies serve neither our needs for basic long-term ecological information nor the more applied and immediate needs of decision making very well.

This brings us to the third institutional problem impeding more effective ecological monitoring. There is currently no national focal point for expertise on ecological problems. There is no one place where government, academia, and industry can come together to address such questions as how to best design studies to detect ecosystem changes, how to develop information bases and systems to maximize existing data, how to improve the capability for analyzing and interpreting impacts, and how to coordinate ecologic research focused on critical issues.

This led the Council on Environmental Quality–Federal Council on Science and Technology in a 1976 report on "The Role of Ecology in Federal Government" to recommend a National Ecological Service—to serve as a focal point for such efforts. Currently, CEQ is leading an Interagency Task Force on Environmental Monitoring and Data, which is continuing to explore this proposal along with many other aspects of the environmental data problem.

In summary, the overall message of this discussion is:

1. Measurement of changes in ecosystem structure and function and analysis of their significance are a very important component of environmental management. We must understand the end result of human actions as contrasted with merely measuring the extent of contamination or a few isolated biologic parameters.
2. We are not doing a good job of ecological monitoring. We are spending many millions nationally on ecological data in response to NEPA and other requirements, but much of this effort is uncoordinated and misdirected. Although measuring ecosystem change is a very tough job scientifically, we are not generally utilizing the state of the art in many of these ongoing efforts.
3. Finally, all this reflects the fact that there are some institutional problems that must be overcome if we are to do better at designing and conducting ecological monitoring programs.

13 National Environmental Research Parks: A Framework for Environmental Health Monitoring

William S. Osburn

For the past ten years I have been involved in a number of meetings having environmental monitoring as their major theme. The scale of actual or proposed monitoring ranged from single cell to ecological monitoring on a global scale. Unfortunately, considerable time was wasted in futile attempts to define and scope biological monitoring and list environmental agents and other ecological stresses that should be monitored.

Why Monitor?

One of the most frustrating questions which had to be addressed was, Why monitor? It is easy to defend biological monitoring programs directed to protecting human health, but to monitor the health of an ecosystem is a program objective that is more difficult to sell to decision-making authorities. Intuitively, many feel that this is a justified area for support while others argue it is absolutely necessary if we wish to maintain a high-quality environment. A few are convinced that human survival depends on an effective ecology monitoring program. However, there are those who insist man is the most sensitive organism and the one to whom we need to focus and apply our monitoring resources.

To indulge in a philosophical quote from Also Leopold that is pertinent to this controversial subject, "A healthful environment begets healthy people." The stated environmental goals of the National Environmental Policy Act (NEPA) of 1969 provides federal agencies with a clear mandate to develop and maintain a comprehensive environmental health monitoring program, and it expresses a philosophy that would be well to adopt on a global basis.

The portion which I feel mandates a full-scale environmental health monitoring program is taken from NEPA, section 101(b):

> ... it is the continuing responsibility of the Federal Government to use all practicable means ... to improve and coordinate federal plans, functions and programs, and resources to the end that the Nation may—
>
> 1. fulfill the responsibilities of each generation as the trustee of the environment for succeeding generations;

2. assure for all Americans safe, healthful, productive and esthetically and culturally pleasing surroundings;
3. attain the widest range of beneficial uses of the environment without degradation, risk to health or safety, or other undesirable and unintended consequences; and
4. preserve important historic, cultural, and natural aspects of our national heritage, and maintain wherever possible an environment which supports diversity and variety of individual choice.

More simply, NEPA tells us to use our environment to achieve a high quality of life but not to abuse it since we are obligated to maintain and preserve for posterity the many environmental alternatives we were so fortunate to inherit. I know of no way to determine the health or condition of these environments without a full-scale environmental monitoring effort.

Thus, the answer to the question of "why monitor?" is no longer philosophical but has the force of federal law behind it.

Before I get into the core of the chapter, I would like to present two analogies which help me frame the problems associated with designing an environmental health monitoring system.

Gradually, over centuries since the days of the Hippocrates, medical doctors, backed by an array of researchers, have developed techniques that provide information for assessing human health. Using indicators such as pulse rate, blood pressure, temperature, and urine analysis, one can obtain a general indication of a person's health. More than twenty-five blood parameters can be determined from a small blood sample; yet, we continually research and attempt to improve these techniques for blood monitoring. For instance, just recently it was reported that a simple procedure had been developed to precisely monitor the amount of oxygen which a prematurely born infant was receiving. Previously, the general skin color or appearance of the baby was used as an indicator that additional oxygen was needed. This nonquantitative approach was fraught with hazard because only slight deviations from the needed levels could be catastrophic.

Millions of dollars have been spent by medical researchers to develop reliable techniques for measuring human health; yet, we know these are in need of improvement. It is possible for a person to drop dead while leaving an examination that pronounced him healthy and sound. If the medical profession has a long way to go before they are out of the woods, ecologists have yet to sight the woods before they have available comparable tools and techniques for measuring ecological health indices.

The second analogy is used to support the contention that any scheme should provide a feedback to improve the monitoring program by reducing costs or improving accuracy.

In the early twentieth century, nearly every rural home canned, stored, or processed food, stored hay and grain enough for their own use and for their livestock to last the winter. Thus, if severe snow storms and/or icestorms could be predicted two or three days in advance, this was sufficient time to take necessary precautions (put cattle in the barn, bring in an extra supply of firewood, and so on) against the storm.

Now let us shift to the present. Think of the dollar value and general convenience of knowing within a few minutes the occurrence, duration, and pathway of a snow storm or icestorm. For instance, crews which sand and salt roadways need to know when, where, and the amount to sand the most hazardous roadways. Miscalculations can be measured in death and injury to people. Now that people are becoming further separated from their base of food and fuel supplies, any disruption from storms is greatly enhanced.

Monitoring can provide basic data needed to plan and can help direct program expenditures for research by enabling one to predict the time, place, and duration of adverse weather patterns.

Present Situation Concerning Monitoring

First I wish to substantiate the contention that there has been a plethora of monitoring conversation with little effort devoted toward developing an integrated national environmental monitoring program. Second, I wish to summarize what the Department of Energy via their National Environmental Research Park program is doing about monitoring the health of our environment.

Despite numerous conferences, workshops, and deliberations of committees in and out of government, we have not established a definitive policy or adequately described national goals or needs. The following statement taken from the 1976 National Research Council report is typical of results: "Establishing a global (or regional) environmental monitoring system for terrestrial ecosystems is exceedingly difficult because of the complexity and diversity of these systems. We were not able to identify a concise set of measurements that could be made on a global basis to fulfill this need." However, most environmental monitoring reports, after noting the difficulty of setting up a monitoring program, present a long list of pollutants which fall into the "should be measured" category.

The recently published "Environmental Monitoring" hearings in September 1977 by the Subcommittee on the Environment and the Atmosphere of the 95th Congress stated: ". . . while much environmental monitoring is being conducted, serious problems exist, especially in terms of data quality, coordination, and comparability, and in the failure to follow through on baseline studies" [1].

This fractionation of monitoring effort among so many is analogous to commissioning individuals to work separately on constructing fuel tanks, carburetors, wheels, brakes, and so on; that is, no amount of superb work on these separate parts will ensure the construction of a car, let alone how it will behave in traffic. These activities must be integrated along the entire assembly line, including a test of the final product.

The Department of Energy (DOE) is presently developing a program for assessing the state of health of regional environments with the intent that these procedures will be generally useful on a national and global basis.

We do know that monitoring the health of the environment will be an expensive and a continuing project. This is why DOE is taking steps to develop prototype protocols for a standard reference system at regional research stations. The first of two projects will be located at the Oak Ridge National Laboratory (Oak Ridge, Tennessee) and at the Pacific Northwest Laboratory (Hanford, Washington). These monitoring programs are part of our National Environmental Research Park program.

National Environmental Research Parks

Very briefly, the concept of a National Environmental Research Park (NERP) developed from a series of field meetings wherein representatives of federal agencies, the academic community, and members of many conservation/environmental groups were invited to Atomic Energy Commission (AEC) facilities in South Carolina, Tennessee, and Washington, D.C., to provide a critical and creative review of the environmental research program. (The NERP charter is reproduced as Appendix 13A.) This was shortly after the National Environmental Pollution Act had been signed; consequently, the spirit of the act strongly influenced this review. A succinct statement of this extensive review was that the field laboratories were "bigger than the AEC" and should be formally considered as national assets. These comments were translated into the form of a tentative research park charter complete with programmatic directives which were tantamount to a national environmental research plan. After several trial drafts, each reviewed by numerous ecological-environmental groups and covering nearly two years' time, a formal charter and program statement was established. It represents the core of this chapter.

The concept of a network of National Environmental Research Parks having as a major purpose "to develop techniques for quantitative and qualitative continuous assessment of the ecological impact of man's activities and technology..." has been well received and was later incorporated as recommendation 8 in the 1974 report, "The Role of Ecology in the Federal Government." Specifically, the recommendation was this: "At present a system of National Environmental Research Parks has been established on certain federally operated facilities. It is recommended that these form the basis for the National System of Ecological Research Areas" [1,2,3].

This chapter deals mostly with the first of the three NERP objectives—to develop methods for quantitatively and continuously assessing and monitoring the environmental impact of human activities.

The National Environmental Research Parks (NERP) program has been established with the following broad objectives: (1) to develop methods for quantitatively and continuously assessing and monitoring the environmental impact of human activities; (2) to develop methods to establish or predict the environmental response to proposed and ongoing activities; and (3) to demonstrate the impact of various activities on the environment and evaluate methods to minimize adverse impacts.

National Environmental Research Parks programs will be different at each site, but a comprehensive research plan for each site will be prepared to ensure that a wide range of research and demonstrations needed to achieve national and agency environmental goals has been adequately addressed. The following specific objectives will serve to guide research for the next few years under the NERP program:

1. Assessment and Monitoring
    a. Compile a regional environmental encyclopedia, including species lists, characterization of ecosystems, successional states, and mapping the vegetation, soils, hydrology, and so on. Species characterizations should include population levels, life histories, and the sensitivity to environmental stresses in terms of behavior, physiology, genetics, reproduction, and productivity. Special attention should be given to endangered and threatened species. This ecological characterization of the region would provide baseline information for superimposing an environmental study and for preparing environmental impact statements.
    b. Set aside and characterize research reference areas. In order to assess the environmental impact of site activities and ecosystem change brought about by stresses, certain minimal representative and/or unique natural areas must be left undisturbed. These reference areas would be established in accordance with other long-range program needs.
    c. Establish field and laboratory repositories. Reference areas should be set aside to serve as gene reservoirs for organisms common to the region and allow a wide range of genetic diversity to be retained. Genetic diversity is a natural defense against the vulnerability to epidemics that tend to develop in plants and animals intensively selected for yield or utility. Sites representing the "regional" deposition of air or waterborne pollutants, not subject to ground-level redistribution, should be preserved and protected from contamination. Repositories of documented environmental samples should be established and maintained as material for determining concentrations of pollutants (including those

yet unacknowledged or unknown) in organisms and the concentration changes over time.
- d. Park sites serve as environmental data centers and ultimately as regional environmental clinics for a particular region of the United States. Since the DOE data storage and retrieval systems are well developed and each site has strong cadres of researchers, DOE has the capacity to set up data centers at each site to store regional information and allow its retrieval, assessment, and dissemination.
- e. Supply basic data so that national environmental decisions, standards, and monitoring programs can be developed with a firm ecological base. A regional monitoring network should be established and operated so that ecosystem responses (including aerobiological systems) can be continually monitored and evaluated with full meteorological information.
- f. Develop and improve ecosystem analysis techniques. Current techniques for conducting ecological surveys, inventorying populations, and measuring ecosystem responses to stresses should be tested and improved, especially those applicable to ecosystems represented at the site.
- g. Manipulate ecosystems, in carefully designed experiments, by applying various environmental stresses and then assessing the ecosystem responses to these perturbations. Techniques should be developed for assessing the general health of ecosystems in terms of their energy flow, materials cycling, species diversity, community structure, and stability.
- h. Participate in international environmental programs to contribute to the assessment of man's global impact on his environment and to cooperate in improving environmental quality.

2. Prediction
   - a. Develop mathematical models simulating ecosystems in order to predict organism response to environmental stimuli, organize knowledge about the system, and select research priorities. The flow of energy, population dynamics, and the cycling of water and nutrients should also be modeled. The accuracy and geographical transferability of these models should be validated.
   - b. Identify organisms, organ systems, and ecosystem components which can serve as indicators of environmental quality.
   - c. Develop techniques to predict the ultimate site and effects of deposition of specific pollutants, particularly those techniques allowing estimation of organism effects without full life history and physiological characterization.
   - d. Study the interactions of pollutants and environmental conditions. Synergistic effects may be very unlike suspected effects of single pollutants, and the prediction of ecosystem and organism response to pollutants depends on the study of interactive effects.

e. Study the pathways and sinks of pollutants in the environment and their sphere of influence. Predicting the ecosystem and organism response to a pollutant entering the environment necessitates determining the ultimate deposition of the pollutant and any other organism or system components which it affects. These studies will complement species and ecosystems characterizations to determine the actual effects of the pollutant.
   f. National Environmental Research Parks serve as sites for successional studies. Environmental insults such as fire, logging, radiation, and thermal effluents continue to impact the environment over time. The rates of secondary succession and/or recovery from such insults should be evaluated and used as a basis for determining the optimal rate of resource exploitation.
3. Demonstration
   a. National Environmental Research Parks serve as public demonstration areas where citizens and specialists can observe long-term effects of specific factors and the true costs of alternate options for waste management or other land uses. Where feasible, National Environmental Research Parks should offer an extension service to community groups, providing lectures, tours, and a visitor's center.
   b. Capitalize on the cadre of trained energy-ecologists and develop a core curriculum training program on-site or in conjunction with a neighboring university.
   c. Demonstrate alternate uses of land. Techniques should be developed for translating ecological costs of energy-related activities into economic ones.

The environmental health monitoring program at Oak Ridge and the Pacific Northwest is in its first year of development. Planning documents including discussions of what, how, when, and where to measure have been prepared. In setting up prototype stations and/or networks, all other groups engaged in environmental monitoring will be contacted.

Much has been learned from others, and procedures as outlined in the Biological Monitoring in UNESCO Biosphere Reserves with special reference to the Great Smoky Mountains National Park will be evaluated and tested. Of specific interest is the stepwise procedure or prediction, monitoring and assessment to test validity of research hypothesis.

References

[1] "Role of Ecology in the Federal Government." Joint Report, CEQ and Federal Council of Science and Technology, Ad hoc Committee Ecological Research, 038-000-00202-7.

[2] *Experimental Ecological Reserves: A Proposed National Network.* Washington: GPO, 1977, stock no. 038-000-00321-6.
[3] "Long-term Ecological Research." Conference Report, Woods Hole, Mass., March 16-18, 1977, National Science Foundation. Copies can be obtained from NTIS, Document Sales, DOC, Springfield, VA 22161.
[4] *Environmental Monitoring,* Hearings before the Subcommittee on the Environment and Atmosphere of the Committee on Science and Technology, U.S. House of Representatives, September 13, 14, 15, 1977, No. 44. Washington: GPO.

Appendix 13A: Charter of the National Environmental Research Parks

A National Environmental Research Park (NERP) is an outdoor laboratory where research may be carried out to achieve national environmental goals, as articulated by the National Environmental Policy Act (NEPA), the Energy Reorganization Act, and the Nonnuclear Energy Research and Development Act. The NEPA translated the public concern for a quality environment into environmental goals, and the National Environmental Research Parks network will provide land to help the Nation comply with the spirit of NEPA. The Energy Reorganization Act of 1974 mandates the Energy Research and Development Administration [ERDA] to engage in environmental research related to the development of energy sources so as to advance the goals of restoring, protecting, and enhancing environmental quality. The National Environmental Research Parks are actually field laboratories set aside for ecological research, for study of the environmental impacts of energy developments, and for informing the public of the environmental and land use options open to them.

Because public access to ERDA land is limited, environmental research projects can be carried out with a minimum of interference. Any land outside Restricted Areas may be made available by the field manager for study under site use procedures. Some natural areas should be protected from all manipulations for definite or indefinite periods of time in order to serve as controls. While the execution of the programmatic missions of ERDA sites must be assured, on-going environmental research projects and protected natural areas must be given careful consideration in any site use decisions. Where appropriate, NERPs may be established with other governmental agencies in interagency agreements.

A wide range of research and demonstration programs will be necessary to systematically address the environmental impacts of man's activities. Research parks are not simply sites to conduct research but have environmental research programs planned to address these general NERP objectives: (1) to develop methods to quantitatively and continuously assess and monitor the environmental impact of man's activities; (2) to develop methods to estimate or predict the environmental response to proposed and on-going activities; and (3) to demonstrate the impact of various activities on the environment and evaluate methods to minimize adverse impacts. NERP programs will be unique at every site, varying in the ecological and energy-related problems addressed and in the participation of researchers which are not funded by ERDA.

14 The National Biological Monitoring Inventory

Robert L. Burgess

The National Biological Monitoring Inventory, initiated in 1975, currently consists of four computerized data bases and voluminous manual files. MAIN BIOMON contains detailed information on 1,021 projects, while MINI BIOMON provides skeletal data for over 3,000 projects in the fifty states, Puerto Rico, the Virgin Islands, plus a few in Canada and Mexico. BIBLIO BIOMON and DIRECTORY BIOMON complete the currently computerized data bases.

The structure of the system provides for on-line search capabilities to generate details of agency sponsorship, indications of funding levels, taxonomic and geographic coverage, length of program life, managerial focus or emphasis, and conditions of the data. Examples of each of these are discussed and illustrated, and potential use of the inventory in a variety of situations is emphasized.

Introduction

There has been a long-recognized need to identify the major monitoring programs in ecology. For many studies of environmental quality, the series data are vitally important. In some cases, the information has already been collected but is not readily available because it has not been published in traditional formats such as professional journals. At times, the data are available only from the investigator. Frequently, the published data that do exist are often years out of date. An important means of tracking these data is by awareness of the programs themselves. Ecological data should be an essential constituent of many environmental quality monitoring programs; however, frequently the pollutant source and transport programs have received most of the attention. The existing ecological monitoring programs have to be inventoried before these gaps can be corrected.

The National Inventory of Selected Biological Monitoring Programs was initiated in June to identify *current or recently completed* biological monitoring projects throughout the United States. Key administrators were identified through a variety of environmental directories (AMA 1973; Clark 1974; CEQ 1973; EPA 1974; FAO/UN 1974; NAS 1974; Paulson 1974; SURC 1971; Thibeau 1972; TIE 1974; Trzyna 1973; Wilson 1974; Wolff 1974). Identification

Research was supported by the Council on Environmental Quality, the Fish and Wildlife Service (USDI), the Department of Energy, and the National Marine Fisheries Service. Publication No. 1279, Environmental Sciences Division, Oak Ridge National Laboratory.

of principal investigators was accomplished primarily by telephone calls to key administrators in all states and to natural resource agencies of local and federal governments. A total of about 7,000 names and addresses of principal investigators was compiled. Computer-generated mailing labels were utilized to send project documentation packages, including questionnaire forms. In addition to a systematic set of code words, enhanced with additional key words provided by the respondents, an abstract, geographical location, data status, statistical treatment, computerization, and availability of data were also requested for each project.

The primary objectives of this inventory were

1. To comprehensively identify and collect information throughout the United States, including continental shelf waters, on biological monitoring studies at the principal investigator/project level
2. To systematically organize the information in computerized files for on-line, interactive searching; for computer projection of reports on technical subject categories, including organisms, study types, management focus, and geographical sites or regions; and to provide complete information retrieval and response/referral services
3. To specifically identify and fully characterize those projects that document changes (that is, time trends, of populations or communities of naturally occurring flora and fauna)

For the purposes of the inventory, we focused on projects that were monitoring natural biota and that demonstrated quantitative change through time. Specifically excluded from the biomonitoring inventory are projects concerned with human population attributes, agriculture, monoculture forestry, domestic animals, economics, Environmental Research and Testing Studies (ERTS), or physical-chemical water quality data where only hydrological, meterological, or physical-chemical water quality data are obtained. Obviously, these criteria also exclude many projects where living organisms are used as indicators of ambient environmental conditions. The inventory has not been concerned with "canary in the mine shaft" types of projects or with those where organisms are effectively used in lieu of physical instrumentation.

Information from the mailings was codified, categorized, computerized, and organized into several data bases. The main biological monitoring data base and supporting data bases are interrelated (figure 14-1). The directory data base contains the names and addresses of principal investigators (about 7,000) to whom the documentation package was mailed. The MINI-BIOMON data base briefly records all responses (more than 3,100) to the inventory. The bibliographic data base contains citations to published documents (about 2,000) received with the responses. The main biomonitoring data base contains over 1,000 selected project responses judged to be the most pertinent and containing the most complete information.

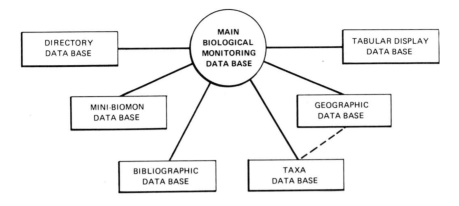

Figure 14-1. The Main and Supporting Biological Monitoring Data Bases

Procedures for developing the remaining data bases (taxa, geographic description, and tabular display) are established, and these will be initiated as time and funds permit. Each will contain more complete project information in selected fields than is contained in the main biological monitoring data base. For example, the geographic-description data base will contain a number of locational descriptor variations that will make it compatible with other geographically oriented systems. Research will be required since much of the desired information is not contained in the responses received to date.

Results

At present over 3,000 responses to the biomonitoring inventory have been received from all states, some U.S. territories, Canada, Mexico, and several countries in the Caribbean (figure 14-2). These can be categorized in a number of ways. By using the documentation form, figure 14-3 summarizes, by numeric total, the numbers of projects in the MAIN data base that utilized the designated terms. By scanning, a general impression of levels of intensity of biomonitoring programs can be gained. For example, item 18 (figure 14-3) indicates that 585 terrestrial, 452 freshwater, and 228 marine projects are currently in inventory files.

An analysis of funding responses for all projects yields a conservative estimate of a total of $126 million per year. This figure is based on the median of the funding-level information requested (less than $10,000, $10,000 to $50,000, $50,000 to $100,000, more than $100,000) and takes no cognizance of the many project responses which contained no funding information. The actual total must be considerably higher, but even an annual budget of $126 million is impressive and

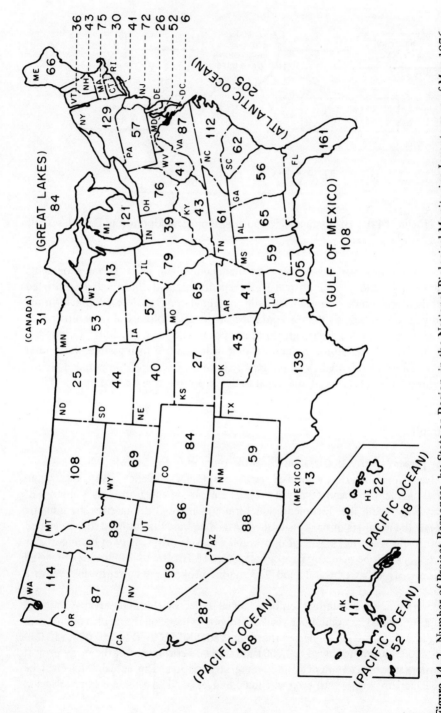

Figure 14-2. Number of Project Responses, by State or Region, in the National Biological Monitoring Inventory as of May 1976

The National Biological Monitoring Inventory

28. (NO. OF SAMPLING SITES)				
No.		or	Other (Specify)	

29. (GEOGRAPHIC DESCRIPTION) (Fill in all applicable descriptions, as known)

Region(s)		State(s)	
County(ies)	1054 names	Longitude	Latitude
River/stream name(s)	522		River mile(s)
Lake name(s)	263	Location other than county(ies)	
Power station name	95	Other industrial site name	
OTHER (Specify)			

30. (GEOCODE UTILIZED) (Circle if applicable)
A. UTM   B. FIPS   C. geodetic (long/lat)   D. grid/map overlay   E. state plane
OTHER (Specify)

31. (PROJECT ABSTRACT) (300 words or less, or attach publication)

In Part B, 880 abstracts (86% of total projects) were provided by responders or adapted by NBIC staff from documents provided by responders. The total number of documents received and stored at NBIC was more than 2000 from the Inventory effort with 694 directly supporting projects contained in the MAIN BIO-MONITORING data base (PART B).

32. (OTHER PROJECTS, OPTIONAL.) (Briefly identify important biomonitoring projects and investigators in your vicinity that are not directly associated with your project)

Project name

Investigator name and address

Project name

Investigator name and address

DESCRIPTION OF DATA RESULTING FROM YOUR PROJECT

33. (DATA TYPE) (Circle all applicable types you retain)
A. raw 798   B. reduced 534   C. summarized 785   D. literature-derived 296
OTHER (Specify)

34. (SUMMARY STATISTICS) (Circle all applicable terms)

A. mean 686   D. percentile 292   G. regression/correlation analysis 393   J. non-parametric analysis 126   M. modeling, stochastic 51
B. mode 141   E. analysis of variance 371   H. contingency table analysis 77   K. non-linear estimation 36   N. serial autocorrelations 21
C. median 178   F. analysis of covariance 109   I. multivariate analysis 164   L. modeling, deterministic 115
OTHER (Specify)

Figure 14-3. The Project Documentation Form with Number of Projects Indicated for the Various Fields

35. (DATA COMPUTERIZED?) In Part B, 250 responders (24% of total) indicated data computerization along with computer type and/or language.		
36. (DATA CODE NAME) (If applicable)		
37. (COMPUTER TYPE)	38. (PROGRAM LANGUAGE)	

39. (DATA AVAILABLE) (Circle applicable term)
   A. unlimited distribution   B. limited to Govt. contract use   C. limited to authorized requesters   D. not available   E. available via publications only
   OTHER (Specify)

40. (DATA AVAILABLE FROM?) (If same as investigator write "same")

Name	Phone No.
Address	
City	State	Zip Code

41. (COST OF DATA TO REQUESTOR) (Circle one)
   A. yes   B. no

42. (SPECIFY UNIT COST AND FORM OF DATA) (Briefly)

43. (PREFERENCE)
   Would you prefer a summary of biological monitoring programs in your State?   Region?   U.S.?   All three?

   Or other? (Specify)

ADDITIONAL INFORMATION

COMMENTS OR REMARKS   Numbers of Federally Sponsored Projects by Agencies:

USDA	94	DOD	37	DOI	187	HEW	8
ARS	9	U.S. Army	3	USGS	5	ERDA	40
FS	74	COE	24	OWRT	23	EPA	46
SCS/CSRS	6	USN/ONR	7	BLM	7	SI	2
DOC	49	USAF	2	BR	8	TVA	6
NOAA	20			FWS	127	NSF	64
NMFS	28			NPS	22	NAS	2
RAPC	1						

NOTE: ANY COMMENTS, REMARKS, OR ADDITIONAL INFORMATION ARE WELCOME.
PLEASE ATTACH AS DESIRED.
REMINDER: We would appreciate receiving any descriptive documents or reports about your project.

Figure 14-3. *Continued*

19. (LEVEL OF STUDY) (Circle all applicable terms)	20. (MANAGEMENT FOCUS) (Circle all applicable terms)
A. organism 541 B. population 790 C. community 578 D. ecosystem 405 E. region 199 OTHER (Specify)	A. endangered species 159 H. range 147 B. environmental impact 566 I. recreation 209 C. fisheries 298 J. resource planning 247 D. wildlife 344 K. pollution control 266 E. forestry 154 L. water quality 316 F. indicator species 367 M. power generation 112 G. radiological 36 N. right-of-way 27 OTHER (Specify)

21. (ORGANISM, MAJOR TYPE) (Circle all appropriate types)

A. Animals 826	2. invertebrate 381	B. Plants 361	C. Microorganisms 272
1. vertebrate 651	a. crustacea (specify order) 248	1. vascular 342	1. viruses 23
a. amphibia 64		a. trees 210	2. bacteria 101
b. reptiles 58		b. shrubs 215	3. fungi 54
c. fish 301	b. insects (specify order) 219	c. herbs 258	4. algae 213
d. birds 236		d. grasses 140	5. protozoa 58
e. mammals 252	c. other (specify phylum)	2. non-vascular (specify below) 50	
	d. mollusca 131	a. mosses 29	
	e. shellfish 67	b. lichens 18	
	f. annelids 66	c. liverworts 3	

OTHER (Specify)

22. (TAXA) (Genus and species, other taxonomic names, or common names if few; or attach listing or publication, if numerous)
There are 832 projects with entries in this field. Of these, 622 contain Latin nomenclature as specified by responders with approximately 2600 genus, species, or other taxonomic names.

23. (KEYWORDS) (In addition to those under other headings. Limit to 10 words)
Approximately 1600 keywords are employed in addition to 132 "code words" employed in items 18, 19, 20, 21, 24, 25, 27, 33, and 34 of the questionnaire.

24. (PRIMARY PARAMETERS STUDIED) (Circle all appropriate types)

					Q. numbers 656	V. sex 232
A. age 351	E. distribution 715	I. habitat characterization 493	M. kill 181	R. physiology 140	W. size/weight 278
B. behavior 245	F. diversity 418	J. harvest 231	N. mortality 405	S. productivity (specify type) 216	X. stress 144
C. biomass 323	G. emigration 179	K. home range size 156	O. migratory patterns 209	T. reproduction 356	Y. territory size 135
D. BOD 54	H. growth 385	L. immigration 165	P. natality 208	U. residues 94	Z. toxicity 100

OTHER (Specify)

25. (SUPPORTING DATA COLLECTED) (Circle all appropriate types)

A. associated fauna 482	C. associated microorganisms 129	E. climatological data 363	G. soils 219	I. water quality data 379
B. associated flora 445	D. atmospheric chemistry 46	F. hydrological data 309	H. terrain 166	

OTHER (Specify)

26. (HOW WERE PRIMARY DATA COLLECTED?) (Use standard terminology, if possible)

27. (SAMPLING FREQUENCY)
A. continuous 135 B. seasonal 347 C. daily 90 D. weekly 182 E. monthly 260 F. quarterly 74 G. annual 246 H. other periodic 166 I. irregular 213

Figure 14-3. *Continued*

PROJECT DOCUMENTATION
NATIONAL INVENTORY OF BIOLOGICAL MONITORING PROGRAMS

PLEASE RETURN COMPLETED FORM TO:

BIOLOGICAL MONITORING INVENTORY
ENVIRONMENTAL SCIENCES DIVISION
OAK RIDGE NATIONAL LABORATORY
OAK RIDGE, TENNESSEE 37830

PHONE: (615) 483-8611
EXT. 3-0391

(Please print in black ink or type)

ADVISORY: THE INSTITUTE OF ECOLOGY

PRINCIPAL INVESTIGATOR

1. (NAME) (Last, First, Initials) Responses to the Inventory total 3192 with approximately the same number of investigators or co-investigators participating.
2. (BUSINESS ADDRESS)
3. (CITY) 4. (STATE) 5. (ZIP)
6. (INVESTIGATOR'S AFFILIATION)
7. (PHONE) ( ) Ext.

DESCRIPTION OF YOUR BIOLOGICAL MONITORING PROJECT
(If you are in charge of more than one project, please use additional enclosed forms as appropriate)

8. (PROJECT TITLE)

9. (PROJECT PURPOSE(S)) (Briefly)

Duration of Funding:
<1 year   43
1–5 years   418
>5 years   493
unknown   53

10. (PROJECT SPONSOR(S))   11. (SPONSOR'S PROJECT OFFICER)
12. (DATE OF INITIATION)
13. (ANNUAL PROJECT FUNDING) (Circle one, or fill in)   OTHER (Specify)
    a. Exact Amount $ b. <$10,000   c. $10-50,000   d. $50-100,000   e. >$100,000
14. (WILL PROJECT BE CONTINUED?) (Circle one)
    a. Yes   b. No   c. Unknown   d. Terminated
15. (MAJOR OBSERVED CHANGE(S) IN TIME) [Briefly describe biotic change(s) identified as a result of your project]

16. (PROGRAM NAME) (If applicable)
17. (PROGRAM DIRECTOR OR ADMINISTRATOR) (If applicable)
18. (ECOSYSTEM OR BIOME TYPE) (Circle all applicable terms)

A. Terrestrial	585	3. savanna	54	B. Freshwater	452	C. Marine	228
1. desert	93	4. grassland	190	1. stream	224	1. coastal	153
2. forest	363	5. tundra	50	2. river	220	2. estuarine	179
a. coniferous	250	a. arctic	32	3. lake	221		
b. deciduous	228	b. alpine	27	4. reservoir	136		
c. tropical	17	6. wetland	131				
OTHER (Specify)		7. mountain	111	Aquatic studies total = 620			

NOTE: We would also appreciate receiving any descriptive documents or reports about your project.

UCN-11687
(3 8-75)

Figure 14-3. *Continued*

indicates the magnitude of the national effort in biomonitoring. A small amount invested in bookkeeping (through this national inventory, for instance) is thus well spent. The projects are sponsored by a diversity of state and federal agencies, teaching institutions, private concerns, and others (table 14-1). The federal government leads in numbers of projects sponsored, while the private sector is poorly represented. While this may reflect an appropriate division of responsibilities, private groups are also the most difficult to identify and inventory.

The information received can also be characterized by sponsorship and taxonomic groups (table 14-2). There are evident disparities between federal and state-level projects, and it appears that different groups or organisms receive more attention in some places than in others. This type of tabular analysis does permit gaps in monitoring coverage to be identified, however, and then adequate planning can address and rectify whatever deficiencies may exist. This concept can be extended to cover agency programs, dollar investment, regional issues, energy technologies, and a variety of potential environmental impacts.

A matrix summary shows the flexibility with which the information can be manipulated and organized, with an example of Atlantic Coast wetland studies (table 14-3). This analysis reflects great interest in endangered and indicator species and in water quality. The degree of interest in power/energy and resource planning also appears to be high.

The matrix format used in table 14-3 can help to make judgments regarding the adequacy of biological monitoring coverage throughout the United States. Care must be exercised in making interpretations of this type, however, because of limitations imposed by our definitions and the degree of coverage achieved. The matrix can be enlarged to dozens of subject categories along each axis, but this is impractical for tabular display purposes. The computer can be used to prepare alternate matrices that may be required for individual states or for regions that can be defined by state boundaries. Further refinement of geographic descriptions (longitude/latitude, country name or code, and so on) will allow more precise summaries.

Table 14-1
Number of Projects in Categories of Funding Sponsors, Based on Responses to the Inventory

Category	Number of Projects
Federal government	1557
State governments	776
Teaching institutions	491
Private concerns	269
Societies and so on	74
Not funded	482

Approximately 9 percent of the projects have multiple sponsors.

Table 14-2
Numbers of Biomonitoring Projects in Each State, Categorized by Funding Source and Sorted on Code Words for Four Major Taxonomic Groups

State	Total Number of Studies	Number of Studies Federally Funded	Number of Studies State-Funded	Number of Bird Studies	Number of Mammal Studies	Number of Fish Studies	Number of Plant Studies
Alabama	66	40	15	1	3	6	4
Alaska	125	79	47	5	22	20	13
Arizona	88	57	21	3	6	7	6
Arkansas	40	23	11	0	0	5	4
California	291	152	43	4	6	30	5
Colorado	84	43	23	4	5	8	8
Connecticut	40	20	10	0	0	5	2
Delaware	27	16	1	0	0	1	0
Florida	162	80	27	3	1	12	10
Georgia	57	32	11	0	1	2	3
Hawaii	16	2	0	0	0	0	0
Idaho	90	51	27	0	6	8	5
Illinois	80	25	28	9	4	8	5
Indiana	38	17	5	1	1	2	3
Iowa	57	18	24	4	3	14	3
Kansas	28	10	8	2	3	6	11
Kentucky	44	18	4	0	0	1	1
Louisiana	103	54	20	2	2	12	2
Maine	67	39	25	2	2	11	7
Maryland	57	36	18	6	4	12	2
Massachusetts	75	37	9	0	0	5	2
Michigan	122	64	26	8	3	13	4
Minnesota	54	34	6	0	1	2	3
Mississippi	60	38	10	2	2	6	2
Missouri	65	32	11	0	0	2	3
Montana	102	67	37	9	14	12	7
Nebraska	40	19	11	5	2	3	0
Nevada	59	37	7	0	2	2	1
New Hampshire	47	25	13	2	1	1	1
New Jersey	73	36	12	4	3	4	3
New Mexico	60	40	9	1	1	1	3
New York	130	56	16	1	0	14	2
North Carolina	112	56	17	1	0	8	6
North Dakota	26	15	9	2	1	2	2
Ohio	78	32	9	2	2	6	4
Oklahoma	43	23	10	0	0	6	1
Oregon	92	63	11	1	3	6	4
Pennsylvania	59	26	8	2	2	3	1
Rhode Island	31	18	6	1	0	4	2
South Carolina	63	33	9	0	0	5	3

Table 14-2 *Continued*

South Dakota	45	29	12	8	4	0	2
Tennessee	61	37	7	2	2	2	2
Texas	143	66	30	11	10	13	11
Utah	86	52	28	6	12	7	10
Vermont	38	24	15	0	7	3	3
Virginia	86	40	11	3	2	4	5
Washington	120	66	25	10	10	7	8
West Virginia	42	22	10	6	4	2	2
Wisconsin	117	43	40	9	5	17	10
Wyoming	70	36	32	9	12	15	8

Totals are meaningless because many projects have joint sponsorship as well as overlap in subject matter and geographical coverage (for example, a single project may cover both birds and mammals in parts of Montana and Wyoming).

Table 14-3
Number of Monitoring Program Responses for Atlantic Coastal Wetlands (126 Total), Arranged by Subject and Management Focus

	Subject Category					
Management Focus	Regional	Site-Specific	Plants	Animals	Micro-organisms	Fisheries
Endangered species	3	15	9	16	3	8
Environmental impact	12	48	28	42	23	22
Indicator species	9	24	12	27	17	15
Power/Energy	8	15	8	18	15	10
Coal	0	0	0	0	0	0
Oil	0	2	0	1	2	0
Nuclear	3	2	2	4	5	2
Right-of-way	3	3	5	6	3	3
Other	2	8	6	7	3	3
Resource planning	6	27	22	26	12	11
Water quality	7	26	21	24	21	17
Total	(17)	(109)	(71)	(87)	(36)	(30)

Numbers in columns are not additive because projects were characterized by multiple use of key words and code words.

Summarized information has been provided to all the initial sponsors (Council on Environmental Quality, Department of Energy, Fish and Wildlife Service, and National Marine Fisheries Service) and also to several offices or laboratories of the Environmental Protection Agency, the Corps of Engineers, the U.S. Geological Survey, the National Oceanic and Atmospheric Administration, the National Park Service, the Nuclear Regulator Commission, and the

National Science Foundation. Exchange of information continues with these and other organizations such as the Arctic Environmental Information and Data Center, Texas System of Natural Laboratories, the Nature Conservancy, Oceanographic Institute of Washington, Cornell University Bird Observatory, Battelle-Columbus Laboratory, and the National Focal Point for the United Nations Environmental Program, International Referral System.

A series of reports have been published, and additional documents are planned. Initially, a study of biological trends indicative of environmental quality changes was prepared from projects identified through the inventory (Suffern et al. 1976). This was an attempt to provide information relevant to the CEQ annual report. The MAIN data base contains full descriptions of 1,021 selected projects (Kemp 1977). More recently, the procedures used to construct the questionnaire have been documented, illustrating how careful preparation and attention to human psychology can result in higher response levels (Kemp et al. 1978).

Selected topics on regional or energy-related themes will be addressed in subsequent documents. Projects will be summarized, data will be analyzed, and interpretive recommendations will be made concerning the status of biological monitoring in these areas of interest.

Summary

The National Inventory of Selected Biological Monitoring Programs presents a source of information for those involved in planning and conducting environmental impact studies, in ecological and ecosystem research, and in all aspects of resource management. Searches of computer files can provide extensive information summaries on individual states or selected regions and a wide variety of technologies. The degree and diversity of responses to the inventory indicate the need for it and its probably future utility. The ability to derive more fully refined information from both the main and supporting data bases will improve as these are developed and supplemented with further information.

References

American Medical Association (AMA). 1973. "A Director of Environmental Organizations in the United States." Chicago: AMA.

Clark, W.E., ed. 1974. *Conservation Directory*. Washington: National Wildlife Federation.

Council on Environmental Quality (CEQ). 1973. *The Federal Environmental Monitoring Directory*. Washington: The President's Council on Environmental Quality.

Environmental Protection Agency (EPA). 1974. *Directory of EPA, State, and Local Environmental Quality Monitoring and Assessment Activities.* Washington: EPA.

Food and Agriculture Organization of the United Nations (FAO/UN). 1974. *Directory of Institutions Engaged in Pollution Investigations.* Rome, Italy: FAO/UN.

Kemp, H.T. 1977. *National Inventory of Selected Biological Monitoring Programs: Summary Report of Current or Recently Completed Projects, 1976.* Oak Ridge, Tenn.: Oak Ridge National Laboratory, ORNL/TM-5792.

———; Goff, F.G., and Ross, J.W. 1978. "Identification of Potential Participant Scientists and Development of Procedures for a National Inventory of Selected Biological Monitoring Programs: A Mail Questionnaire Survey." Oak Ridge, Tenn.: Oak Ridge National Laboratory, ORNL/TM-6224.

National Academy of Sciences (NAS). 1974. *Internatinal Biological Program Directory.* Washington: National Academy of Sciences, National Academy of Engineering, National Research Council.

Paulson, G., ed. 1974. *Environment USA.: A Guide to Agencies, People, and Resources.* New York: R.R. Bowker Co.

Suffern, J. Samuel; West, D.C.; Kemp, H.T.; and Burgess, R.L. 1976. "Biological Monitoring and Selected Trends in Environmental Quality." Oak Ridge, Tenn.: Oak Ridge National Laboratory, ORNL/TM-5606.

Syracuse University Research Corporation (SURC). 1971. *Environmental Research Laboratories in the Federal Government.* New York: SURC.

The Institute of Ecology (TIE). 1974. *Directory of Environmental Life Scientists,* vol. 1 to 9. Washington: TIE.

Thibeau, C.E., ed. 1972. *Directory of Environmental Information Sources.* 2d ed. Boston: The National Foundation for Environmental Control, Inc.

Trzyna, T.C., ed. 1973. *World Directory of Environmental Organizations.* San Francisco: Sierra Club.

Wilson, W.K., ed. 1974. *World Directory of Environmental Research Centers.* 2d ed. New York: R.R. Bowker Co.

Wolff, G.R., ed. 1974. *Environmental Information Sources Handbook.* N.Y.: Garwood R. Wolff Co.

Part V
Bioassay of Environmental Pollution

The development of rapid, effective, and inexpensive means to detect and monitor toxic contaminants in the environment are described in part V. Irreversible biological effects of greatest societal concern include cancer, birth defects, and heritable genetic changes. Authors describe various bioassay methods available to monitor these potential effects of toxic contaminants by using selective, reproducible indicator organisms.

The Ames *Salmonella* microsome test for mutagenesis is cited as the cornerstone in a battery of short-term bioassays from prokaryotic microorganisms to human cells in culture.

Chemical mutagens have been identified in a broad range of products, such as pharmaceuticals, cosmetics, food additives, agricultural chemicals, and in industrial effluents. Research using plant systems to monitor environmental mutagens is described.

Behavioral and neurological changes are useful for early detection and in measuring the presence and effects of toxic agents.

15 Short-Term Bioassays of Environmental Samples

Michael D. Waters and
Linda W. Little

Introduction

The protection of human health and environmental quality depends on detection and prevention of release of harmful agents. With more than 50,000 chemicals in common use and countless combinations of agents in manmade waste products, it is clear that conventional approaches in environmental surveillance are inadequate. The development of rapid, effective, and inexpensive means for evaluating the health and environmental impact of toxic agents is critically needed. Short-term tests for health and ecological effects offer promise in detecting toxic and genotoxic materials and in prioritizing these substances for further examination by established procedures. The potential health effects of greatest societal concern include cancer, birth defects, and heritable genetic change. The Ames *Salmonella* microsome test for mutagenesis has provided the cornerstone for a battery of short-term bioassays of progressively greater complexity from prokaryotic microorganisms to human cells in culture (Ames, Lee, and Durston 1973). Potential ecological effects of concern include stimulation or inhibition of the growth and/or biological activity of primary producers such as algae in the aquatic environment; stress to primary producers such as soybeans in the terrestrial environment; and toxicity to key aquatic organisms such as fish and shrimp. This chapter considers various approaches in the application of the available bioassays to problems of environmental research and monitoring. Emphasis is placed on the type of biological activity detected (that is, the role of each test within the battery), implementation status, complexity, cost, and sample requirements.

A Developing Environmental Assessment Program

Because of the urgency in dealing in a cost-effective manner with the environmental impact of chemicals presently in use or proposed for use, the Environmental Protection Agency has delineated a number of proposed testing schemes. The first of these to be implemented in pilot studies is the environmental assessment program of the Industrial Environmental Research Laboratory, Research Triangle Park, North Carolina (IERL-RTP), which has delineated a three-phased approach to performing an environmental source assessment (that is, the evaluation of feed and effluent streams of industrial processes in order to determine the need for control technology) (Hamersma, Reynolds, and Maddalone 1976; Duke, Davis, and Dennis 1977; Dorsey et al. 1977). This approach includes short-term tests for potential health effects as well as tests for aquatic and terrestrial ecological effects. The first-level testing is designed to provide preliminary assessment data, help identify problem areas, and generate information needed to determine priority of pollutants or pollutant sources for additional testing.

First-level biological testing is conducted in parallel with chemical analysis, since it is clear that these disciplines have a dual role in the process of environmental assessment. Chemical analysis cannot provide sufficient data for complete evaluation of potential pollutant effects because the biological activity of complex samples cannot be consistently predicted. However, although bioassays can indicate the biological activity of a given sample, they cannot specify which components of a crude sample are responsible for the observed biological effects. A cost-effective screening approach therefore involves the use of short-term bioassays to determine which samples are biologically active together with chemical fractionation and analysis to ascertain which agents are responsible for the observed effects.

Proposed Level 2 biotesting, though not definite at this time, will provide additional information to confirm and expand information gathered in Level 1 testing. Level 3 testing will involve determination of the toxic or inhibitory components in a waste stream as a function of time and process variation, along with monitoring of any chronic, sublethal biological effects.

The IERL-RTP environmental assessment program, while still under development, represents an important beginning in the assessment of environmental hazards based on the use of short-term bioassays. Clearly, however, short-term tests do not circumvent the need for conventional toxicological, clinical, and epidemiological evaluation, nor are they intended to do so. They do indicate which pollutants and pollutant sources should receive high priority for these long-term and comparatively expensive procedures.

Biotesting of Samples for Potential Health Effects

There is now available a matrix of short-term health effects bioassays which can be applied in a phased or stepwise manner in the biological analysis of environmental samples. It should be noted, however, that some short-term bioassays are known to be insensitive to specific chemicals or classes of chemicals. No single bioassay is adequate for monitoring all types of chemical and biological activity. In the health effects area, this problem may be addressed by the use of a required "core" battery of tests and identification, before biotesting, of those samples containing substances similar in structure or activity to those known to give "false negative" results in the test.

In the area of mutagenic and carcinogenic testing, several variations of the tiered, multilevel, or phased approach have been discussed in the literature (Flamm 1974; Bridges 1976; Dean 1976). Overall, there is substantial agreement on the point of emphasis at each evaluation level and on the need for a battery of tests.

A phased approach to bioscreening for environmental health effects emphasizing kinds of bioassays is illustrated in figure 15-1. This is a three-step matrix with a battery of tests at each level. The emphasis in the phase 1 test battery is on the *detection* of acute toxicity by using mammalian cells in culture and intact animals, genotoxic effects including point mutation and primary DNA damage in microbial species, and chromosomal alternations in mammalian cells in culture. The Phase 2 battery is designed to *verify* the results from Phase 1 tests by employing higher-level toxicity tests involving mammalian cells in culture and intact mammals and genotoxicity assays using plants, insects, and mammals. Genotoxicity assays at Phase 2 are separated into tests for mutagenesis per se and specific tests for carcinogenic potential. Phase 3 testing is devoted to quantitative *risk assessment,* using conventional toxicological methods. For the purpose of defining a probably negative result of genotoxicity, the "core" battery of short-term tests is most important.

"Core" Battery of Tests for Genotoxic Effects

Since no single test is capable of indicating all the various types of biological activity which may be relevant to the processes of mutagenesis and carcinogenesis, it is generally held that a battery of short-term tests should be performed. The battery approach is intended to minimize false negatives and thus ensure protection of human health. Batteries of tests have been proposed in the development of the EPA Pesticide Guidelines for Mutagenicity Testing and in the Consumer Product Safety Commission's "Principles and Procedures for Evaluating the Toxicity of Household Substances" (NAS 1977). These documents reflect the thinking expressed in the Committee 17 report on environmental mutagenic hazards (Drake 1975) and in the report of the working group of the DHEW Subcommittee on Environmental Mutagenesis (Flamm 1977). Indeed, there is considerable agreement that a core battery of tests for mutagenic and carcinogenic effects should include, as a minimum, tests for point mutation in microorganisms and gene mutation in mammalian cells in cultures; tests for chromosomal alterations, preferably an in vivo test; a test for primary damage to DNA using mammalian, preferably human, cells in culture; and a test for oncogenic transformation in vitro. Such a battery of tests might be considered to represent the "core," or the most essential, of the genotoxicity tests in the phased evaluation process. Redundancy in the test battery is considered desirable until a more complete data base of test results can be assembled. Also, to aid in

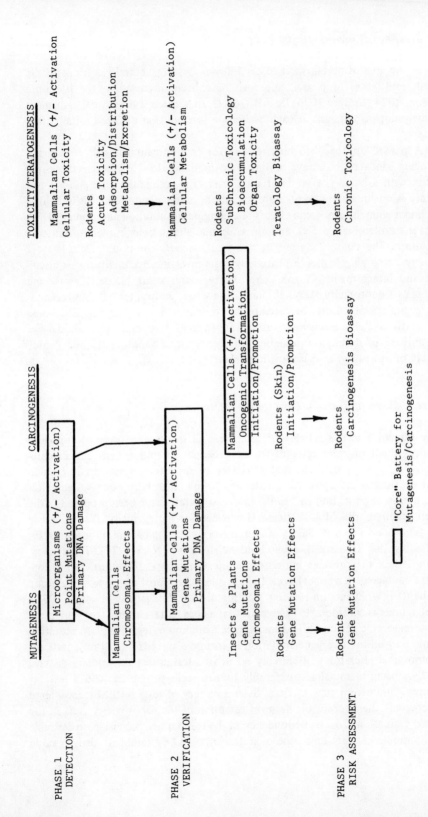

Figure 15-1. A Phased Approach to Bioscreening for Environmental Health Effects

interpretation, it is necessary to ascertain the influence of cellular toxicity in these tests. The following is a description of the kinds of biological activity detectable in short-term tests comprising the core battery.

Biological Activity Detected in the Core Battery

Point or Gene Mutation. Point mutations are alterations which affect single genes. These alterations include base pair substitutions and frame-shift mutations, as well as other small deletions and insertions. Applicable test systems include both forward and reverse mutation assays in bacteria, yeast, and mammalian cells in culture. Most of these assays employ an exogenous source of metabolic activation provided by a mammalian liver microsomal preparation. It has become apparent that a majority of the genotoxins are procarcinogens or promutagens which must be converted into their reactive forms before their effects can be evaluated. The metabolic conversion is believed to be mediated by oxidative enzymes and to involve the formation of reactive electrophillic metabolites which bind covalently to DNA. Gene mutation assays which incorporate whole-animal metabolic activation (for example, urine screening) are very desirable since it is not possible to ensure metabolic fidelity in entirely in vitro systems. Furthermore, one must employ intact animals to demonstrate the heritability of mutational effects.

Chromosomal Alterations. Chromosomal alterations include the loss or gain of entire chromosomes, chromosome breaks, nondisjunctions, and translocations. Tests for these abnormalities involve searching for chromosomal aberrations in somatic and germinal cells, usually from insects and mammals. Chromosomal aberrations observed in germinal tissues of intact animals provide important evidence of the accessibility of the test chemical to the reproductive organs.

Primary Damage and Repair in Informational Macromolecules. Damage and repair bioassays do not measure mutation directly but do measure the direct damage to DNA and other macromolecules by chemical agents and its subsequent repair. Bioassays to detect macromolecular damage and repair are available which use bacteria, yeast, mammalian cells, and whole animals. Except for the whole-animal bioassays, these systems generally employ an exogenous source of metabolic activation.

Oncogenic Transformation in Vitro. Oncogenic transformation is the process whereby normal cells grown in culture are converted into malignant cells after treatment with a carcinogen. The demonstration of malignancy (tumor formation) can be observed by injecting the transformed cells into whole animals, although this is not a requisite for oncogenic transformation. A number of mammalian oncogenic transformation bioassays utilizing cells derived from different rodent species are available. Some of these cell systems have the endogenous capability to activate procarcinogens while with others exogenous microsomal activation have been used successfully.

Toxicity in Vitro. An initial requirement in mammalian cell mutagenesis and oncogenesis bioassays is the determination of the lethal toxicity of each test agent. This information may be used to establish the range of concentrations to be employed in the mutagenesis or oncogenesis assays and to quantify the observed mutation or transformation frequency in terms of the number of cells surviving the treatment.

Utility as Predictive Tools. Two critical questions are: How good are these genotoxicity tests? and What do they mean?

Carcinogenesis. At present, microbial mutagenesis test systems are used most widely to prescreen substances for potential oncogenicity. Tests for gene or point mutations in microorganisms, as, for example, those involving the *Salmonella typhimurium* microsome system, have been found to be highly predictive of oncogenic potential (McCann and Ames 1976). Indeed, most chemical mutagens which have been adequately tested have been found to be oncogenic in whole-animal bioassays. It is well established that most, but not all, oncogens are mutagens when appropriate metabolic activation is provided in the short-term tests. Research on test systems which permit sequential evaluation of mutagenesis and oncogenic transformation has enhanced our understanding of the relationship between these two phenomena.

Oncogenic transformation in vitro is considered to be directly relevant to the process of tumor formation in the intact animal. Few, if any, false positives are detected by using this methodology (Pienta, Poiley, and Lebherz 1977). However, because of the laborious nature of cell transformation assays, it may not be feasible, because of time or fiscal constraints or the availability of facilities, to immediately put a very large number of samples through such testing procedures. It is for reasons such as these that oncogenic transformation assays are generally considered to be higher-level tests.

When more fully developed, human cell bioassays for oncogenic transformation may afford a final short-term test for substances found positive in phylogenetically lower organisms. This will be true especially if epithelial cell systems can be developed which retain their metabolic activation capability. Those chemicals which produce positive responses in human cell systems might be given highest priority for evaluation in conventional whole-animal oncogenesis bioassays.

Mutagenesis. The fundamental concern in mutagenic testing is the risk to future generations. Alterations of the genetic material in germinal cells are rarely expressed in the exposed individuals. These alterations may not become apparent for several generations, but they contribute to an increased genetic burden. The observation and quantitation of these mutational effects in germinal tissues require the use of intact animals (for example, the sex-linked recessive lethal test in the insect *Drosophila*) (Wurgler, Sobels, and Govel 1977). It would be highly desirable to include such a test in the core battery if significant human exposure is anticipated to a suspect mutagen.

The usefulness of cells other than germinal cells as a predictive tool is judged to be high for certain kinds of genetic alterations. Microbial cells are widely used to detect point mutagens. Many of these systems have geen "genetically engineered" to enhance their sensitivity as detection systems. Mammalian and human cells in culture can provide more relevant information on the ability of the substance to induce both point mutations and chromosomal alterations. It is important to evaluate both types of potential genetic activity. Chromosomal alterations are best evaluated in intact animals, but inexpensive, whole-animal tests for gene mutagens are lacking.

Ancillary Effects. Tests which detect primary damage to informational cellular macromolecules (say, DNA) have been found to show moderate to high correlation with mutagenic and oncogenic potential as indicated by animal bioassays (Sugimura et al. 1976; Energy Resources Co. 1977). Microbial tests in this category are extremely rapid and inexpensive and offer the possibility of examining various manifestations of macromolicular damage. Mammalian and human cell culture bioassays for DNA damage offer the possibility of detections with mutagenic and oncogenic potential as evaluated by using experimental animals (San and Stich 1975). Several of these systems permit concomitant measurement of primary DNA damage, point mutation, chromosomal effects, and cellular toxicity.

Strategy for the Employment of a Phased Approach

As indicated previously, a phased or stepwise mode of application has been favored as a cost-effective approach to the bioscreening of large numbers of complex environmental samples and their components.

A strategy for the employment of short-term bioassays based on biological activity, cost, and complexity may be delineated as follows:

In Phase 1, tests representing each kind of biological activity would be performed. The extent of redundancy in testing within a category of biological activity would be dictated by a number of factors, including production volume, anticipated human exposure, known hazards of related samples, and so forth. If any of these tests proved positive, the appropriate follow-up tests would be pursued in Phase 2 and, if required, in Phase 3 depending on the degree of associated risk. If those tests were to prove negative, no further testing would be performed unless there were overriding considerations, as mentioned previously. In such cases a core battery of tests for genotoxicity would be completed with negative results before short-term testing would cease. Extensive health risk could entail further long-term testing to define a negative result. This approach would facilitate a cost-effective utilization of limited testing resources and at the same time provide protection for human health in proportion to the anticipated risk involved.

Specific tests may be organized within the test matrix as follows:

Phase 1 Bioassays. A battery of short-term tests for this phase is illustrated in table 15-1. As mentioned previously, the emphasis at this level of testing is on *detection* of mutagens, potential carcinogens, and acutely toxic chemicals in a battery of in vitro and in vivo tests. The results obtained from Phase 1 tests are used to assign priorities for further testing in appropriate confirmatory bioassays at Phase 2. The in vitro endpoints considered, based on expense, complexity, and the current level of development of bioassay systems, are point mutations, chromosomal alterations, primary DNA damage, and cellular toxicity. All bioassays are performed with and without mammalian metabolic activation systems where feasible. Conventional rodent acute-toxicity tests are considered essential at Phase 1 in view of the limitations of cytotoxicity screening tests. The latter

**Table 15-1
Phase 1 Short-Term Bioassays for Mutagenesis/Carcinogensis/Toxicity**

Point Mutations
Salmonella microsome (Ames) reverse mutation, prototrophy to histidine
Escherichia coli–WP2 microsome reverse mutation, prototrophy to tryptophane
Saccharomyces cerevisiae reverse and forward mutation

Chromosomal Effects
In vitro cytogenetics: Chinese hamster ovary cells; WI-38 human fibroblasts

Primary DNA Damage
Escherichia coli Pol A$^-$ repair deficient strains
Bacillus subtillis, REC$^-$ repair deficient strains
Saccharomyces cerevisiae mitotic recombination

Cytotoxicity
Rabbit alveolar macrophage (for particulates)
Chinese hamster ovary cells
WI-38 human lung fibroblasts

Rodent Acute Toxicity

on-line tests cannot represent intact animals but provide useful preliminary information about the relative cellular toxicity of selected samples, for example, airborne particulate materials (Waters, Huisingh, and Garrett 1978). In addition, rodent acute-toxicity tests can provide a source of body fluids and tissues to be examined for the presence of active mutagens and carcinogens by the use of short-term genotoxicity bioassays.

Phase 2 Bioassays. As mentioned previously, Phase 2 tests are designed to verify the results obtained in Phase 1. Examples of Phase 2 tests are shown in table 15-2. The test systems are selected to provide confirmatory information on gene mutations, chromosomal alterations, primary DNA damage and repair, and cellular oncogenic transformation. The latter test provides more explicit information on the carcinogenic potential of a sample. The test organisms are mammalian cells in culture supplemented with exogenous metabolic activation, plants, insects, and intact mammals. These systems are considered to provide more relevant and definitive information in the continuing process of health hazard evaluation, especially where intact organisms are employed.

Table 5-2
Phase 2 Short-Term Bioassays for Mutagenesis/Carcinogenesis/Toxicity/Neurospora

Gene Mutations
Mammalian cells in culture (CHO, L5178Y, V79)
Insects—*Drosophila*
Plants—*Tradescantia* and maize

Chromosomal Effects
In vivo cytogenetics—Leucocyte culture and bone marrow cells

Primary DNA Damage
Unscheduled DNA synthesis (WI-38)
Sister-chromated exchange formation (in vitro and in vivo)

Oncogenic Transformation
Syrian hamster embryo cells
Mouse fibroblast cell lines (C3H10T1/2 and BALB/c 3T3)

Cellular Metabolism
Primary liver cells

Rodent Subchronic Toxicology

Teratology

Phase 3 Bioassays. Phase 3 testing involves the use of conventional whole-animal methods. The emphasis here is on quantitative risk assessment. Experimentation with intact mammals is needed to provide infromation on the presence, concentration, and biological activity of toxins in the target issues. In addition, information on pharmacokinetics involving absorption, distribution, metabolic transformation, and excretion cannot be obtained without studies using intact mammals.

Resource Implications. The resource implications of a mutagenicity/carcinogenicity bioscreening program are shown in table 15-3. It is evident that most

Table 15-3
Resource Implications of Carcinogenicity/Mutagenicity Bioscreening Program

Tests	Cost[b] ($)	Study Time[c]	Quantity of Material Required (g)rams
Gene (Point) Mutations			
Bacteria (Ames plate test)	350– 600	2– 4 weeks	2
Bacteria (liquid suspension)	1,000– 2,000	2– 4 weeks	2
Eukaryotic microorganisms (yeast)	200– 500	2– 4 weeks	2
Neurospora	2,000– 3,000	4– 6 weeks	2
Insects (*Drosophila*, recessive lethal)	6,000– 7,500	4– 6 months	10
Mammalian somatic cells in culture (mouse lymphoma)	2,500– 4,800	1– 2 months	2
Mouse-specific locus	20,000+	1 year	25
Chromosomal Alterations[a]			
In vitro cytogenetics	1,000– 2,000	2– 4 weeks	2
In vivo cytogenetics	3,000– 6,500	6– 8 weeks	20
Insects, heritable chromosomal effects	3,000– 6,500	4– 6 weeks	10
(*Drosophila*), nondisjunction	3,000	1– 3 months	
Dominant lethal in rodents	6,000–10,000	3 months	20–25
Heritable translocation in rodents	40,000–67,000	12–18 months	25
Primary DNA Damage			
DNA repair in bacteria	200– 500	2– 4 weeks	2
Unscheduled DNA synthesis	350– 2,000	4– 6 weeks	2–5
Mitotic recombination in yeast	200– 500	4– 6 weeks	2–5
Sister chromatid exchange	1,000– 1,200	4– 6 weeks	2–5
Oncogenic Transformation in Vitro			
Chemically induced transformation	6,500– 7,500	10–12 weeks	2–5

[a]These tests may be classified as tests for chromosomal alterations.
[b]The cost of these tests has varied and can be expected to vary until test requirements are stabilized.
[c]This period covers the experimental time and report preparation.

of the short-term bioassays designed to detect point mutation, chromosomal alterations in vitro, and primary DNA damage and repair are relatively rapid and inexpensive and require small amounts of test material. Together with cytotoxicity bioassays for selected applications and rodent acute-toxicity tests, these bioassays constitute effective screens for toxic and genotoxic effects of energy-related emissions and effluents.

Biotesting of Samples for Potential Ecological Effects

Level 1 environmental assessment ecological testing delineated by IERL-RTP also includes short-term tests for assessing impact of pollutants on nonhuman species, especially those representing important components of aquatic and terrestrial communities. Again, these tests are designed to provide the *first step* of a phased approach and are not intended as substitutes for such definitive tests as long-term or chronic tests at sublethal toxicant concentrations.

Aquatic Tests

Recent increased emphasis on aquatic bioassays is dramatically illustrated by changes in a basic reference in the field, *Standard Methods for the Examination of Water and Wastewater* (APHA 1976). The 13th edition (1970) contained about fifty pages on bioassay methods, emphasizing the fish bioassay procedure, while the 14th edition includes about 200 pages of assays with a wide variety of organisms. With increased interest, substantial changes in terminology have also occurred in the direction of more specificity. For this reason, one must consult recent texts for currently accepted terminology and must exercise care in interpreting results presented in older research publications. As an example, the terms LD_{50} (the median lethal dose of a toxicant), TL_{50} (the median tolerance limit), and LC_{50} (median lethal concentration) were used almost interchangeably in the past in relation to aquatic bioassays. Strictly speaking, the term *dose* refers to the amount of toxicant that actually enters the test organism, a parameter seldom measured in aquatic assays. The term "TL_{50}" does not specify what effect was measured. For these reasons, the term *concentration*, referring to the concentration of toxicant in the test solution, is now preferred. Where lethality is the endpoint, the term "LC_{50}" is used; for effects other than mortality, the term "EC_{50}" (Median effective concentration), along with a statement of the exact effect noted, is employed.

The Level 1 ecological effects tests are shown in table 15-4 and are described in detail in the IERL-RTP manual (Duke, Davis, and Dennis 1977). Descriptions of aquatic tests are also available in other publications (APHA 1976; Peltier 1978; Duthie 1977; and others).

Table 15-4
Level 1 Short-Term Bioassays for Ecological Effects Assessment

Aquatic
Freshwater algal assay procedure: Bottle test
Acute static bioassays with freshwater fish and *Daphnia*
Marine algal assay test
Static bioassays with marine fish and shrimp

Terrestrial
Stress ethylene plant-response test
Soil microcosm test

Algae are important primary producers in aquatic ecosystems, serving as the basis for the food chain and as contributors to reaeration. If overabundant, however, they form the undesirable "blooms" associated with eutrophication. The algal assay tests are designed to measure either stimulatory or inhibitory effects of pollutants. They were originally designed to determine the potential of waste waters or other substances, such as detergents, to stimulate growth of algae by increasing the available amount of some limiting nutrient such as nitrogen or phosphorus. Essentially, the tests involve inoculation with algae of a control medium and media containing various concentrations of the test material, followed by measurement of growth by cell counts, dry weight, turbidity, or other methods for about two weeks. The recommended freshwater test organisms are the green alga *Selenastrum capricornutum,* the blue-green algae *Microcystis aeruginosa* and *Anabaena flos-aquae,* and the diatoms *Cyclotella* and *Nitschia*. The recommended marine alga is *Skeletonema costatum*. The tests require about 14 days of incubation in addition to a week or so for preparation of inocula. The freshwater assays cost approximately $150 to $300 and require about 1.2 labor days per sample tested. The marine assay costs approximately $500 and requires about the same amount of time.

Probably, because effects of pollutants on fish can be highly visible to the public, fish toxicity tests have a long history. The acute static fish bioassay, one of the oldest tests in the field of aquatic toxicology, has been the target of many jests, is frequently called "pickle jar ecology" because of the 5-gal jars generally used as test containers, and has been criticized as being over simplistic. Nevertheless, its value, especially in screening, has been amply demonstrated. The test generally involves exposure of a minimum of ten fish in each of a series of test concentrations, in dilution water of known quality, for 96 hr at constant temperature. The results are expressed in terms of the LC_{50}. The test requires a total of 4 days, assuming a stock of acclimated fish. The test costs approximately $400 and requires approximately 1 day of labor per sample tested. Fish employed in the test are generally the freshwater fathead minnow, *Pimephales promelas,* and the marine sheepshead minnow.

The test may also be used for invertebrates such as the freshwater *Daphnia* of the marine grass shrimp. These tests require less space, media, and test sample.

As noted previously, these bioassays may detect effects at concentrations beneath analytical capability to measure, and they may alert one to the presence of some toxicant for which one had not thought to analyze.

Terrestrial Tests

In comparison to the aquatic bioassay procedures, the terrestrial procedures are very new and relatively untried. They were included in Level 1 testing because of the dearth of short-term tests in this area and the desire to include some tests for assessment of potential impact of pollutants on terrestrial species. Because of recognized problems with these procedures when they are used for routine screening, it is difficult to assign costs to them. The IERL-RTP test manual (Duke, Davis, and Dennis 1977) describes the tests in detail and includes references to the work on which they are based.

The stress ethylene plant-response test used for air pollutants is based on response of plants to environmental stress by release of elevated levels of ethylene. The recommended plant is the soybean *Glycine max*. The plants must be grown in chambers with controlled temperature and lighting and be exposed under controlled conditions to at least three levels of sample gas. Ethylene levels in the ambient atmosphere are determined by gas chromatography.

The soil-microcosm test is designed to assess (1) the mobility of the test substance and its degradation products, (2) the primary mode of soil export (gaseous, dissolved, or particulate), (3) effects on nutrient cycling, (4) effects on soil biota, and (5) dose-response effects on the above. The test employs cores of soil taken from representative terrestrial sites, periodically leached with rainwater or other dilution water before (3 weeks) and after (3 weeks) the addition of the test material to the surface of the soil.

Projected Level 2 ecological effects bioassays will include bioaccumulation and persistence studies, and possible Level 3 assays include mutagenicity, ecosystems analysis, and waste treatability tests.

Conclusions

No single bioassay is entirely adequate for assessing the impact of a substance on human health or ecological systems. Even a battery of short-term tests cannot provide a definitive assessment of toxicity, mutagenicity, or other effects. Nevertheless, a multilevel approach with batteries of tests at each level, conducted in parallel with chemical analysis, offers a cost-effective and time-effective alternative to the critical problem of evaluating the impact of toxic agents on human health and the environment.

References

American Public Health Association 1976. (APHA). *Standard Methods for the Examination of Water and Wastewater,* 14th ed. Washington: APHA.

Ames, B.N.; Lee, F.D.; Durston, W.E. 1973. An improved bacterial test system for the detection and classification of mutagens and carcinogens. *Proceedings of the National Academy of Sciences, USA* 71:782-786.

Bridges, B.A. 1976. "Use of a Three-Tier Protocol for Evaluation of Long-Term Toxic Hazards Particularly Mutagenicity and Carcinogenicity." In *Screening Tests in Chemical Carcinogenesis,* eds. R. Montesaro, H. Bartsch, and L. Tomatis. WHO/IARC Pub. No. 12, Lyon, France, pp. 549-568.

Dean, B.J. 1976. A predictive testing scheme for carcinogenicity and mutagenicity of industrial chemicals. *Mutation Research* 41:83-88.

Dorsey, J.A.; Johnson, L.D.; Statnick, R.M.; and Lochmuller, C.H. 1977. *Environmental Assessment Sampling and Analysis: Phased Approach and Techniques for Level 1.* Washington: EPA, EPA 600/2-77-115.

Drake, J.W. 1975. Environmental mutagenic hazards. (Prepared by Committee 17 of the Environmental Mutagen Society. *Science* 187:504-514.

Duke, K.M.; Davis, M.E.; and Dennis, A.J. 1977. "IERL-RTP Procedures Manual: Level 1 Environmental Assessment Biological Tests for Pilot Studies." Washington: EPA, EPA 600/7-77-43.

Duthie, J.R. 1977. "The Importance of Sequential Assessment in Test Programs for Evaluating Hazard to Aquatic Life." In *Aquatic Toxicology and Hazard Evaluation* (ASTM STP 634) eds. F.L. Mayer and J.L. Hamelink. Philadelphia: American Society for Testing and Materials, pp. 17-35.

Energy Resources Co. 1977. "Short-Term Toxicological Bioassays and Their Applicability to EPA Regulatory Decision Making for Pesticides and Toxic Substances." Report prepared for the Office of Planning and Evaluation by Energy Resources Co., Inc., EPA Contract Number 68-01-4383.

Flamm, W.G. 1974. A tier system approach to mutagen testing. *Mutation Research* 26:329-333.

_____. 1977. "Approaches to Determining the Mutagenic Properties of Chemicals: Risk to Future Generations." Prepared for the DHEW Committee to Coordinate Toxicology and Related Programs by working group of the Subcommittee on Environmental Mutagenesis.

Hamersma, J.W.; Reynolds, S.L.; and Maddalone, R.F. 1976. "IERL-RTP Procedures Manual: Level 1 Environmental Assessment." Washington: EPA, EPA 600/2-76-160a.

McCann, J., and Ames, B.N. 1976. Detection of carcinogens as mutagens in the *Salmonella*/microsome test: Assay of 300 chemicals. *Proceedings of the National Academy of Sciences USA* 73:950-954.

National Academy of Sciences (NAS). 1977. "Principles and Procedures for Evaluating the Toxicity of Household Substances." Prepared for the Consumer Product Safety Commission by the Committee for the Revision of NAS Publication 1138. Washington: National Academy of Sciences, pp. 86-98.

Peltier, W. 1978. *Methods for Measuring the Acute Toxicity of Effluents to Aquatic Organisms*. Washington: EPA, EPA 600/4-78-012.

Pienta, R.J., Poiley, J.A.; and Lebherz, W.B. 1977. Morphological transformation of early passage golden Syrian hamster embryo cells derived from cryopreserved primary cultures as a reliable *in vitro* bioassay for identifying diverse carcinogens. *International Journal of Cancer* 19:642-655.

San, R.H.C., and Stich, H.F. 1975. DNA repair synthesis of cultured human cells as a rapid bioassay for chemical carcinogens. *International Journal of Cancer* 16:284-291.

Sugimura, T.; Sato, S.; Nagao, M.; Yshagi, T.; Mutsushima, T.; Seino, Y.; Takeuchi, M.; and Kawachi, T. 1976. "Overlapping of Carcinogens and Mutagens." In *Fundamentals in Cancer Prevention*, eds. P.N. Magee et al. Baltimore, Md.: University Park Press, pp. 1919-215.

Waters, M.D., and Epler, J.L. 1978. "Status of Bioscreening of Emissions and Effluents from Energy Technologies." Third National Conference on the Interagency Energy/Environment R&D Program, Washington, June 1-2, 1978. In *Proceedings of Energy/Environment III*, pp. 29-59. Springfield, Va.: NTIS, EPA 600/9/78-002.

_____; Huisingh, J.L; and Garrett, N.E. 1978. "The Cellular Toxicity of Complex Environmental Mixtures." Symposium on Application of Short-term Bioassays in the Fractionation and Analysis of Complex Environmental Mixtures, Williamsburg, Va. *Proceedings* in press.

Wurgler, F.E.; Sobels, F.H.; and Govel, E. 1977. "*Drosophilia* as Assay System for Detecting Genetic Changes." In *Handbook of Mutagenicity Test Procedures*, eds. B.J. Kilbey, M. Legator, W. Nichols, and C. Ramel. Amsterdam: Elsevier Science Publishing Co., pp. 335-373.

16 Plants as Monitors for Environmental Mutagens

Michael D. Shelby

Current widespread concern over the risk to human welfare posed by chemical mutagens in the environment is based on three facts: (1) humans are exposed to chemicals, (2) chemicals can cause mutations, and (3) mutations can cause disease. Although there are no confirmed instances of chemically induced genetic disease in humans, a significant portion of human disease is known to have a genetic basis. In hereditary diseases such as hemophilia, sickle cell anemia, and phenylketonuria, the patterns of inheritance are straightforward and well understood. In other cases, such as diabetes, hypertension, and schizophrenia, there is evidence for transmission from generation to generation, but the patterns of inheritance are not clear and therefore the genetic bases are obscure. It is not known what proportion of today's genetic disease results from preexisting mutations and what proportion from mutations recently induced by mutagenic factors in the environment. There is, nevertheless, a sound basis for believing that exposure to mutagens can increase the incidence of human genetic disease. A thorough discussion of the problem of environmental mutagens can be found in a recently published report of the Subcommittee on Environmental Mutagenesis, DHEW Committee to Coordinate Toxicology and Related Programs [1].

Chemicals which are capable of inducing mutations have been identified in a broad range of structural classes and are found in many different usage categories including pharmaceuticals, cosmetics, food additives, pesticides, and other compounds in a host of industrial processes and products. The air, water, and soil are all components of the human environment which have become polluted with mutagens and many other chemicals as a result of human activities— activities which, to different degrees, are beneficial to man. The occurrence of mutagens in major components of the environment puts large segments of the human population at risk of exposure to potentially hazardous chemicals. Since such chemicals do pose a hazard and can be controlled, it is important that we develop and utilize means by which to monitor their occurrence.

Because mutagens interact with DNA to produce mutations and DNA is the hereditary material of all living things, any organism is a potential tool in the study of mutagenesis. To be suitable as a monitor for mutagens occurring in the environment, an organism should require a minimum of preparation and maintenance. It should be well understood genetically and provide an easily detected endpoint for mutation induction. The assay system should be highly sensitive since environmental exposures are likely to be at very low doses. The detection of small increases in mutation frequencies will furthermore require reliable control data.

With these considerations in mind, plants are a group of organisms which seem well suited for a monitoring role. In order to evaluate higher plant systems with potential as monitors for mutagens and to define the role they may play in environmental monitoring the National Institute of Environmental Health Sciences sponsored a workshop in January 1978 entitled "Higher Plant Systems as Monitors of Environmental Mutagens." Twenty-three papers were presented at this workshop, and the manuscripts were published in *Environmental Health Perspectives*, volume 27.

Among the most promising systems discussed at the workshop was the *Tradescantia* stamen hair system developed at Brookhaven National Laboratory. In this sytem mutagenic events are detected as color changes from blue to pink in the cells of stamen hairs [2]. The stamen hair system has proved to be a highly reliable and sensitive assay system in radiation studies over the past three decades. Recent studies have extended its application to chemical mutagenesis, particularly in the detection of gaseous mutagens under controlled laboratory conditions. The stamen hair system recently has been coupled with a specially designed mobile laboratory for the detection of mutagenic activity in ambient air.

The new mobile monitoring laboratory is equipped with controlled-growth chambers where *Tradescantia* cuttings are maintained. In one chamber, plants are exposed to ambient air while concurrent controls in another chamber are exposed to filtered air. The mobile unit is taken to selected sites where plants are exposed for ten days. Plants are then returned to the laboratory where frequencies of mutational events in the stamen hairs are determined. Pilot studies using the mobile laboratory have been carried out at several sites in the United States and have yielded promising results in the ability of this sytem to detect low levels of mutagenic activity in air pollutants [3].

Another monitoring system under investigation involves the use of corn (*Zea mays*) to detect mutagens in the soil. Current emphasis has been placed on the detection of mutagenic activity resulting from pesticides. The detection system is based on staining the pollen with iodine. Plants homozygous for a mutant waxy allele (*wx*) are grown on soil to which a pesticide is applied. Such plants normally produce haploid pollen grains, which contain the recessive *wx* allele and do not stain with iodine. When there is a mutation back to the wild-type allele (*wx*), however, the mutated pollen grains stain blue-black. These can easily be scored against a background of unstained pollen by using a dissecting microscope. Several pesticides, applied at commonly used concentrations, have been shown to increase the frequency of mutations at the waxy locus [4]. Experiments designed to further evaluate the potential of this system as an in situ monitor are in progress. A highlight of this work has been its bearing on the question of whether plants have the ability to metabolically convert promutagens to their active forms. Recent data from the waxy system indicate that such metabolic conversion does occur [5]. In further studies using microbial indicators, positive results were also obtained with extracts from pesticide-treated corn plants and more recently with a mixture containing the pesticide and an extract from untreated plants [6]. The pesticide gave negative results when tested alone in microbial assays. The nature of the active metabolites as well as the enzyme systems leading to activation is under investigation. Although they are preliminary, results from the waxy assay as well as those obtained with microbial indicators point to mutagens activated by the metabolic systems of plants as another possible source of human mutagen exposure. Thus there is a need for further research in this area.

The use of natural populations of plants as monitors for mutagens in the environment is an appealing concept which also has been explored [7]. Although there are problems with establishing proper controls and with the general paucity of genetic knowledge of such populations, the idea deserves further investigation. Recent work with populations of the royal fern, *Osmunda regalis,* has revealed unusually high levels of chromosomal damage in populations growing near a river polluted with pulpwood industry wastes [8]. The lack of such damage in upland populations has led to speculation that the damage may have been induced by mutagenic pollutants in the river. While further work is needed to confirm such conclusions, these findings point out the potential usefulness of ferns and other natural plant populations as in situ mutagen monitors. A complete discussion of the fern system can be found in Klekowski [9].

A number of other mutagen assay systems have been developed in plants and have been used extensively in mutagenicity studies of both radiation and chemicals [10, 11]. Assays available in plants such as soybean (*Glycine max*), barley (*Hordeum vulgare*), and *Arabidopsis* and a variety of systems for detecting chromosome aberrations may hold potential for development as monitoring systems, but little or no work has been carried out to evaluate their suitability for this purpose [12, 11, 13, 14].

In summary, the development and application of plant genetic systems to provide sensitive and inexpensive mutagen monitors is a new but promising area of research in environmental mutagenesis. If developed and employed, such systems could yield, in addition to an early detection of mutagens, information valuable to epidemiologists concerned with the occurrence of genetic disease. In view of the correlation between the mutagenicity and carcinogenicity of chemicals, mutagen monitoring data might also provide key information in our understanding of the geographic patterns of cancer mortality as revealed in recently published cancer atlases of the United States [15, 16].

References

[1] Drake, John W.; Abrahamson, Seymour; Crow, James F.; Hollaender, Alexander; Lederberg, Seymour; Legator, Marvin S.; Neel, James V.; Shaw, Margery W.; Sutton, H. Eldon; von Borstel, R.C.; Zimmering, Stanley; de

Serres, Frederick J.; and Flamm, W. Gary. Approaches to determining the mutagenic properties of chemicals: Risk to future generations. *Journal of Environmental Pathology and Toxicology* 1:301-352 (1977).

[2] Underbrink, A.G.; Schairer, L.A.; and Sparrow, A.H. *Tradescantia Stamen Hairs: A Radiobiological Test System Applicable to Chemical Mutagenesis in Chemical Mutagens,* vol. 3, ed. A. Hollaender. New York: Plenum Press, 1973.

[3] Schairer, L.A.; Van't Hof, J.; Hayes, C.G.; Burton, R.M.; and de Serres, F.J. Exploratory monitoring of air pollutants for mutagenic activity with the *Tradescantia* stamen hair system. *Environmental Health Perspectives* 27:51-67 (1978).

[4] Plewa, M.J., and Gentile, J.M. Plant activation of herbicides into mutagens—The mutagenicity of field-applied atrazine on maize germ cells. *Mutation Research* 38:390 (1976).

[5] Gentile, J.M., and Plewa, M.J. Plant activation of herbicides into mutagens—The mutagenicity of atrazine metabolites in maize kernels. *Mutation Research* 38:390-391 (1976).

[6] Plewa, M.J. Activation of chemicals into mutagens by green plants: A preliminary discussion. *Environmental Health Perspectives* 27:45-50 (1978).

[7] Tomkins, D.J., and Grant, W.F. Monitoring natural vegetation for herbicide-induced chromosomal aberrations. *Mutation Research* 36:73-84 (1976).

[8] Klekowski, E.J., Jr., and Berger, B.B. Chromosome mutations in a fern population growing in a polluted environment: A bioassay for mutagens in aquatic environments. *American Journal of Botany* 63:239-246 (1976).

[9] Klekowski, E.J. "Detection of Mutational Damage in Fern Populations: An *in situ* Bioassay for Mutagens in Aquatic Ecosystems." In *Chemical Mutagens: Principles and Methods for Their Detection,* vol. 5, ed. A. Hollaender. New York: Plenum Press, 1978.

[10] Ehrenberg, L. "Higher Plants." In *Chemical Mutagens: Principles and Methods for Their Detection,* vol. 3, ed. A. Hollaender. New York: Plenum Press, 1973.

[11] Nilan, R.A., and Vig, B.K. "Plant Test Systems for Detection of Chemical Mutagens." In *Chemical Mutagens: Principles and Methods of Their Detection,* vol. 4, ed. A. Hollaender. New York: Plenum Press, 1973.

[12] Vig, B.K. Soybean (*Glycine max*): A new test system for study of genetic parameters as affected by environmental mutagens. *Mutation Research* 31:49-56 (1975).

[13] Redei, G.P. *Arabidopsis* as a genetic tool. *Annual Review of Genetics* 9:111-128 (1975).

[14] Grant, W.F. Chromosome aberrations in plants as a monitoring system. *Environmental Health Perspectives* 27:37-43 (1978).

[15] Mason, T.J.; McKay, F.W.; Hoover, R.; *Atlas of Cancer Mortality for U.S. Countries: 1950-1969*. Washington: Department of Health, Education, and Welfare, Public Health Service, 1975, (NIH) 75-780.

[16] Mason, T.J.; McKay, F.W.; Hoover, R.; Blot, W.L.; Fraumeric, Jr. J.F. *Atlas of Cancer Mortality among U.S. Nonwhites: 1950-1969*. Washington: Department of Health, Education, and Welfare, Public Health Service, 1976. (NIH) 76-1204.

17 The Use of Behavioral Techniques in Biological Monitoring Programs

C.L. Mitchell

In the past the United States and Western European countries have placed the major emphasis in studies of the toxicity of environmental agents on defining the effects of these agents in terms of pathological changes with special emphasis on carcinogenicity. It is becoming more and more apparent, however, that a focus on the functional aspects of toxicity is at least as important as pathological aspects. This is particularly true with respect to the nervous system where nonconcordance between pathology and function is frequently observed. Indeed, the functional capacity of the central nervous system to compensate for focal cortical damage is well documented. Conversely, gross behavioral impairments may not be accompanied by discernible central-nervous-system lesions on postmortem examination.

Two interrelated methodological problems which confront behavioral and neurological toxicology as it applies to environmental agents are in the insidious onset of effects and the (frequently) subjective nature of the complaints, at least early in the toxic process. Because of these problems, there is limited agreement as to the sensitivity and utility of many commonly used behavioral tests and procedures, when applied to the study of environmental toxic agent. Thus, one of the research goals of the Laboratory of Behavioral and Neurological Toxicology at the National Institute of Environmental Health Sciences is to develop a strategy for the selection of behavioral and neurological procedures for the study of environmental toxicants. It is our view that little meaningful progress can be made in behavioral and neurological toxicology until this has been accomplished.

We most certainly are not faced with a shortage of available methods; rather, we are faced with a relative lack of rational criteria with which to choose suitable methods. This, in my opinion, is the major problem with using behavioral techniques in biological monitoring programs. Certainly, chemists do not want to use an instrument which has not been calibrated properly or which is not sensitive to the agent being monitored. Similarly, behavioral toxicologists need to standardize and validate their methods before they can be maximally utilized in monitoring programs.

In any monitoring program, one has to define the thing for which one is monitoring. This is especially true when behavioral methods are used. The high degree of integration of the nervous system precludes using a single test as an indicator for all possible effects on the system. Thus, the most rational approach will involve the use of method(s) known to be sensitive to the particular agent being monitored.

It should also be emphasized that with any suitable biological method the power of that test to detect an effect at a given concentration of the toxic substance is a function of the size of the biological sample. Thus, the selection of the appropriate sample size is an integral part of any biological monitoring system.

Part VI
Critical Issues

The following workshop agenda was prepared to take advantage of the varied backgrounds and scientific interests of participants in biological monitoring and to suggest various issues needing resolutions.

Five workshop groups met to consider these topics and to make recommendations on four broad areas. The chairman and/or synthesizer summarized the group's findings before a plenary session, and final recommendations were prepared after further communication with workshop participants following this workshop.

1. Communications and Dissemination of Information
   R. Burgess, Chairman; James Stewart, Synthesizer
   a. Need for an information center with computer-assigned capability to abstract and inventory biological monitoring research, legislation, facilities, and scientists
   b. Need for a society to represent scientists actively engaged in work on biological monitoring and provide a continuing forum to discuss problems and progress and to promote the technology
   c. Possible need of a national technical publication
2. Research Needs and Priorities
   J.D. Buffington, Chairman; Linda Little, Synthesizer
   a. Assessment of scientific validity
   b. Criteria for selecting indicator organisms
3. Regulatory Control and Public Policy
   C. Weber, Chairman
   a. Federal, state, and local legislation requiring data on biological effects from in situ biological monitoring or laboratory bioassay
   b. Biological monitoring of occupational environments
   c. Biological monitoring of industrial and municipal effluents
4. Methodological Issues in Biological Monitoring
   A. Maki, Chairman (Short-term Effects); Ken Dickson, Chairman; Dan Dindal, Synthesizer (Long-term Effects)
   a. In situ monitoring and the development of standardized methods
   b. Bioassay method to determine short- and long-term biological effects
   c. Methods to combine physical-chemical and biological techniques

18 Communications and Dissemination of Information

Robert L. Burgess and
James Stewart

The committee addressed a series of items connected with the general problem of communication and information dissemination. The potentials for either a single information center or a series of these were discussed, and the recommendation was made that a single center should have the responsibility for inventory collation, aggregation of information, and dissemination. A recommendation was also made that the current effort at Oak Ridge National Laboratory for the past three years be continued and that the efforts be updated and expanded to serve a relatively broad user group.

A second topic was the problem of a need for society or an organization to coordinate the various groups and individuals interested in all the facets of biological monitoring. Such an organization could serve the interests of the public, scientific groups, regulatory officials, and managers at all levels. First, we considered the formation of a society which could serve as a focal point for the various needs of those concerned with biological monitoring. Second, as an alternative to that, we would like to consider the possibilities of forming, from existing organizations, some type of consortium, that is, committees, sections, or subgroups within the various societies that have an interest in some area of biological monitoring. This conceptual approach should provide an effective group with common objectives under which we can begin to operate in a biological monitoring context. Many groups that were mentioned included various kinds of microbiological societies, ecological societies, chemical societies, soil societies, and so on. Ideas regarding the formation of a society on the involvement of working groups within an existing society should be forwarded to the committee chairman. Efforts will be made to canvass the interest in a group of societies that can be identified in order to explore the potential for a meeting of representatives of these to investigate the potential for a consortium.

The third item considered was the need for a national or other kind of publication. The consensus was that this was something that needed further exploration. We discussed the possibilities of putting together a somewhat general document, perhaps a newsletter, that would address primarily the public groups. The biological monitoring group represented at this conference comprises a rather broad array of societies, all of whom have journals, newsletters, or bulletins of their own, and it was suggested that efforts be made to incorporate into those publications information such as short write-ups of the results of this workshop.

Finally, a fourth item was deliberated—the need to make information and research needs of biological monitoring more available. One recommendation concerned the possibilities of holding regional biological monitoring workshops across the country. It was also suggested that a national meeting be held following this series of regional workshops to look closely and from different perspectives at many of the problems addressed here.

The other aspect of public information discussed was the already-stated need for certain kinds of publications that would, in fact, interpret the uses for and the nature of biological monitoring for a wider public group. This would involve a type of public relations activity. It has been said that biological monitoring has arrived and will burgeon in the next few years, based largely on national and international needs. Public support is needed to elicit government action and support. The user public should be identified and invited to future activities of this type. Although an important step has been made in this workshop, it was felt that there was a significant segment of the biological community that could well have profited from this experience.

In summary, a recommendation was made for a very active approach to be made to a broad group of users or potential users or interested individuals and agencies, rather than the somewhat more passive approach of putting out a notice to determine interests.

19 Research Needs and Priorities

J.D. Buffington and *L.W. Little*

Research and educational needs can be ascertained only after thorough consideration of some basic questions: What are the goals of biomonitoring? What are we trying to measure? Why?

Biological testing, including biomonitoring, has generally been divided into two categories: human health effects testing and ecological effects testing. Traditionally, research development and applications in these two areas have occurred in totally different milieus, but consensus of the workshop participants was that there should be more interaction and integration. Such interaction is especially important in view of the trend toward "tier testing" or "test battery" approaches in evaluating the impact of chemicals released into the environment.

Workshop participants felt that certain aspects of biomonitoring research warranted special consideration:

1. Experimental design, data collection, data interpretation
2. Case histories and predictive models
3. Problems in distinguishing between natural variations and anthropogenerated changes
4. Institutional arrangements for interdisciplinary research

Experimental Design, Data Collection, and Data Interpretation

The opinion was expressed that too much funding goes into gathering massive collections of data which are filed away, never to be examined again (or maybe never examined at all). There needs to be continuous feedback into the experimental design process so that appropriate data will be collected in future studies. It often appears that a lot of "answers" have been amassed without any advance attempt to formulate relevant questions.

In biomonitoring, identification of organisms through the species level is required. In this regard, participants felt that there is a dire need for better taxonomic aids and for well-trained taxonomists. Aids such as reference slides, keys, and culture-type collections would be useful. The difficulty in getting funding for taxonomic work was pointed out. If someone goes in with a proposal for conducting research on identification of species in some phylum, he will probably not get funded unless he can tie this into a larger monitoring project.

Another point of discussion was, How should data be presented in order to influence environmental decisions? Data that have been collected and interpreted usually will be reinterpreted by a "user" who is not necessarily trained in assessment of its value. The user may be an administrator not trained in science at all but who holds the purse strings or otherwise influences what is going to be funded, who is going to fund it, who is going to do it, and so on. Furthermore, information is generally transmitted to someone in the lower echelons in an agency; from there it goes through a chain of command up to the administrative level. Any person in that chain of command who misunderstands the data can interfere with the whole evaluation process.

Case Histories and Predictive Models

Before industrial facilities, including nuclear generation facilities, are built, very detailed, operational, voluminous environmental impact statements (EIS) must be filed. In the EIS, predictions of the impact of the operation on the environment are included. An effort should be made *after* the plant is operating to ascertain if, indeed, the predictions made were correct and whether the massive data collected were, in fact, relevant to the questions to be answered. One must ask, Will this operation have an adverse effect on the environment, and if so, what and to what extent? Case studies would be especially useful in indicating the elasticity and resilience of ecosystems in response to stress.

Workshop participants felt that research should be directed toward utilizing the information already available to design better predictive models.

Distinguishing between Natural Variations and Changes Caused by Humans

Biomonitoring may be addressed to short-term or long-term changes. In either case, it is necessary to determine if something is happening and if so, what the significance of this change? Are both the biotic and abiotic components of ecosystems undergoing natural variations in response to known and unknown stimuli. These variations may occur in a diel, seasonal, or longer-term pattern, and they contribute to the "background noise" when one is attempting to assess the significance of a change in biological activity, especially if the change is small.

Obviously, knowledge of natural variations is necessary in order to determine the significance of the biotic responses measured in biomonitoring.

The amount of information in this general area is exceedingly sparse and much needed to intelligently design and interpret biomonitoring projects. To obtain the information would be a tedious, long-term process requiring the combined efforts of scientists from many disciplines. These factors may make obtaining research funding difficult and will, at the very least, require a reeducation of funding agencies under pressure to get a lot of "numbers" in a short time.

Institutional Arrangements for Interdisciplinary Research

Almost by definition, biomonitoring has to be a multidisciplinary team approach. Yet, because of academic department and government agency organization, it is often very difficult to assemble such a team and deal with the administrative aspects. How is the money to be distributed? Who makes decisions as to how the project is coordinated? Participants felt that in most colleges and universities there is little interaction between members of different departments. The "research center" approach might be as one way to simplify organizational arrangements and encourage team approaches.

Participants also felt that there should be more communication within and between federal agencies. The different branches of an agency often are not cognizant of what other branches of the same agency are doing. As a consequence, several agencies may be funding similar projects in one research area while no funds are available for other important areas.

Other points discussed in the workshop sessions include:

1. The need for biomonitors in indoor environments
2. The need for microbial indicators of change
3. The utility of benthic macroinvertebrates in biomonitoring
4. The possibility of developing a bioassay procedure by using microorganisms to assess groundwater quality
5. The ideal components of the educational program of "environmental toxicologists" and "environmental health scientists"

20 Regulatory Control and Public Policy
C. Weber

General Recommendation

Inasmuch as the *principal* objective of the federal Environmental Protection Program is to restore and maintain the biological integrity of the environment, it necessarily follows that if no adverse biological environmental effects are noted (including effects on human and other terrestrial and aquatic life), we can assume that the biological integrity of the environment is indeed being maintained. In other words, it can be said that "If you don't have a biological problem in the environment, you don't have a problem." If biological effects are the principal effects of concern, biological monitoring is the most direct and efficacious form of monitoring and therefore should have the highest priority within the EPA monitoring program (that is, should have precedence over chemical monitoring). It should also receive the highest priority in funding.

Recommendation Relating to Effluent Monitoring

The EPA should explore the possibility of using a single, simplified, unified approach to measuring the biological properties of effluents (and/or emissions), possibly with microorganisms, which can be used to monitor pollution in water, air, and human work environments.

Recommendations Relating to Receiving-Water Monitoring

Federal agencies should allocate more resources to solve problems in assessing the effects of pollutants on receiving waters in order to

1. Develop protocols for evaluating the effects of specific pollutants (problem-solving approach), such as sediments, petrochemicals, and so on.
2. Develop defined, quantitative, maximum-acceptable limits of "impairment" of community structure and function that would be permitted by state and federal regulations.
3. Develop "norms" or "standards" of community structure and function with which to compare field data collected in biomonitoring programs.
4. Define what is meant by "fishable" waters.
5. Standardize methods of data evaluation.

21 Methods Issues in Biological Monitoring— Short-Term Organization of Biological Monitoring Methods

Alan W. Maki

Introduction

A workshop session was convened on the morning of March 23, 1978, consisting of workshop participants interested in the numerous and varied issues of methodology for biological monitoring. Because of the large size of the group (approaching fifty individuals), it was decided to split the session to achieve a more workable group size, thus affording all participants the opportunity to express their individual viewpoints on issues of concern within the limited time. To effect the split, the group was halved along the somewhat artificial distinction between those individuals interested in methods issues for long- versus short-term biological monitoring programs. Operationally, this arbitrary distinction between long- and short-term programs was made along the lines of those interested in monitoring methods for predictive toxicology with studies of mammals and aquatic life, environmental spill impacts, and general acute-impact studies, while those interested in long-term applications considered subtle aquatic and terrestrial community shifts, continuous monitoring of point- and nonpoint-source effluent discharges, and methods for monitoring the long-term impact of industrial and energy-related facilities.

Those interested in long-term methodology were organized under the chairmanship of Dr. Kenneth L. Dickson, and the deliberations of that group are the subject of a separate summary. The first half of the group, then, focused on methods issues concerned with short-term monitoring programs, and the following represents a summary of those discussions.

The initial approach taken was to elicit from the participants an agenda of basic issues and problems currently confronting methods application in biological monitoring. Following the generation of this issues list, the specific points were prioritized by the group, and discussions proceeded to deal with each throughout the morning session. The outline agenda for these issues is presented.

Use and Application of Biological Monitoring

A discussion was organized to elicit from all participants specific examples of either planned or operational biological monitoring programs. It was thought that a listing and discussion of these programs would exemplify numerous applications of biological monitoring in use and thus address the basic questions of why the respective program managers were applying or considering the various methods. That is, why is biological monitoring useful?

The Pros and Cons of Standardized Methods for Biological Monitoring

The focus of this session was a consideration of the relative merits of standardized methods for biological monitoring programs. Considering the numerous techniques evolving and myriad of applications, will the development of standard methods significantly enhance the application of results to real-world systems? What gaps currently exist between methods employed and real-world applications? Will an overall approach to standardization significantly enhance quality assurance programs or cost-effectiveness? What considerations should be given to the ultimate selection, culture, and care of test species to be employed in biological monitoring programs? What criteria can be established to enhance the legal defensibility and overall acceptability of resultant monitoring data?

Specific Decision Criteria Indicating the Need for Biological Monitoring

Do common data requirements exist during laboratory and field ecological investigation which specifically key or identify the need for biological monitoring programs? Can a common list of decision criteria be generated and employed to aid investigators in identifying the need for and timely application of biological monitoring methods? Do specific chemical and physical characterizations of the test material or specific environment in question imply the need for biological monitoring? How can an investigator identify these needs early in his research efforts?

The Development of Regional, State, and Federal Policies for Biological Monitoring

In recognition of rapidly increasing interest in the development of biological monitoring methods, is it advisable at this early stage of development to initiate consistent regulatory policies to direct the evolution and application of biological monitoring methods? If so, what should be the general substance of these policy recommendations and at what governmental level should these policies arise?

Short-Term Organization 205

Discussion and Conclusions

Use and Applications of Biological Monitoring

Each participant in this workshop session was asked to relate his experiences with the development, use, and application of biological monitoring programs in his particular field of research interest. The discussions were to include a consideration of why biological monitoring was chosen from among the options available to the investigators, and how the data was to be utilized to further the overall objectives of the study design.

As anticipated, a wide spectrum of diverse and interesting applications of biological monitoring methods resulted from the twenty session participants. Since several methods had generally similar objectives and approaches applied to various environmental questions, the resulting examples of programs are grouped into five separate categories of use.

Effluent Testing. Several participants expressed an interest in biological methods for the continuous monitoring of point-source effluent quality. Methods employed included on-site, continuous-flow toxicity testing with effluent dilutions, test species entrained in chambers receiving effluent, and automated methods for enumeration of coliform bacteria as an index of ambient sewage concentrations. The reasons for applying these methods ranged from requirements under National Pollution Discharge Elimination System (NPDES) permit stipulations to justifications made on a cost-effectiveness basis. When compared to indepth fish population surveys in receiving waters, toxicity tests with effluent dilutions afford a reasonable and relevant alternative.

Toxicity Testing. Several individuals were employing biomonitoring methods with representative aquatic species for short-term testing in a predictive context. By monitoring subtle physiological changes within the test individuals, those concentrations eliciting effects were correlated with existing data from long-term chronic studies to assess the predictive value of the short-term monitoring approach. Time and costs savings were obvious criteria dictating the development of these methods. Current methods for chronic toxicity evaluation are expensive and time-consuming because of the requirement to expose the test individuals throughout their entire life cycles. Monitoring and subsequent predictions of chronic toxicity from observed physiological effects yield a substantial improvement by offering a predictive screening tool for use earlier in the program for evaluation of overall hazard of a new substance.

Similar desires to employ an early screening tool in a predictive context were voiced by several participants in their decisions to utilize the Ames test for evaluation of effects of chemical substances on mammalian systems. However, it was recognized by the group that these promising methods for predictive toxicology require a substantially larger data base with many more test materials before generalized recommendations for application of these short-term methods can be made with a reasonable degree of confidence.

Environmental Impact Studies. Although the balance of studies discussed during this session ultimately relates to environmental impact studies of some specific nature, several participants listed long-term, multidisciplinary monitoring studies as an integrated unit to evaluate environmental impact of manufacturing or energy-related operations. Monitoring methodology is employed to evaluate benthic community response and recovery as a result of construction programs— location of waste-water treatment plants, power generating stations, and dredging or channelization programs. Recent technology in oil-shale mining and coal gasification have spawned the need for receiving-water evaluations of resultant mine wastes, runoff, and wastes generated from the processes themselves. These monitoring methods have focused on the development of monitoring methods for the actual sampling of structure and function of receiving-water communities. Ranging from in-stream sampling of fish and macroinvertebrate communities to field evaluations of phytosynthesis/respiration (P/R) ratios, these methods have the advantage (and assocated problems) of sampling the actual community in question and providing the opportunity for detailed analysis of subtle shifts in structure and function of the real-world ecosystem.

Several described their monitoring activities to determine the structure and health of local plant species relevant to information needed in an evaluation of ambient air quality. Stationary plants continuously exposed to airborne pollutants offer distinct advantages for evaluation of long-term effects associated with specific emissions. Similarly, methods are evolving for the monitoring of airborne microorganisms in an effort to understand disease and/or pathogen transmissions. As with methods for monitoring the quality of the aquatic environment, the overall utility and cost-effectiveness of these methods must be assessed on an individual basis, and additional programs are needed to more fully assess the practical utility of these approaches.

Regulatory Interests. A high degree of interest was shown by several participants in the application of monitoring methods to compliance monitoring of effluent quality and for programs focused on the registration of new chemicals under recent state and federal legislation. Several state water quality control boards and EPA Region 4 currently have specific biological monitoring requirements for determination of effluent quality on a relatively routine basis. Although these methods generally focus on relatively simple and rapid acute-toxicity screens, in many instances they do serve as the basis for initial determination and enforcement of discharge permit stipulations.

During the conduct of integrated programs designed to provide data for the registration or clearance of new chemicals, biological monitoring offers several distinct advantages, ranging from predictive toxicological testing in the laboratory to effluent assessments from manufacturing facilities and resultant effects in receiving waters following consumer use and disposal patterns. Individual decisions to employ these monitoring programs for effects testing must be based on a rational scheme designed to provide relevant data while avoiding the generation of superfluous testing data or information of little relevance to the overall safety evaluation program.

Biological Monitoring Methods as a Teaching Aid. Participants affiliated with academic institutions expressed strong interest in the application of biological monitoring methods as classroom or laboratory aids to describe the structure, function, and assimilative capacity of aquatic and terrestrial ecosystems in response to a chemical, physical, or biological perturbation. Laboratory model ecosystems offer the size and convenient complexity for simple and rapid quantification of effects and interactions on the model community with some degree of prediction to real-world incidents. Courses in the environmental sciences have much to gain from an application of laboratory and field methods of biological monitoring. They offer the student an opportunity to observe, first hand, species and community function and the impact of test materials or environmental variables on their normal functions. Equipping a mobile trailer with a basic wet laboratory was mentioned as a possibility for basic limnology field courses, offering students the opportunity to perform field analyses by establishing a regular biomonitoring program for particular aquatic environments within the immediate area.

The Pros and Cons of Standardized Methods for Biological Monitoring

One of the first issues confronting the investigator who is considering the implementation of a biological monitoring program, either as an adjunct to an existing program or as a separate and unique investigation, is the selection of appropriate test species offering the greatest potential for meaningful data acquisition. The workshop participants felt that a discussion of criteria which lead to and define test species selection would be useful to individuals considering monitoring programs by both providing direction to their respective programs and providing advance consideration of potential problems in the culture, care, and maintenance of the test species throughout the experimental period. The criteria for biomonitoring test species selection and their rationale are listed.

Objectives of the Study. The participants felt that attempts to overly standardize test species for particular experimentation would inevitably lead to a loss of ecological meaning and relevance for many programs. The group, therefore, recommended against the adoption of rigidly standardized test species, instead leaving the option to the experimental design and particular objectives of the study.

Ease of Culture and Maintenance. An important criterion in the ultimate selection of test species, after the particular study objectives are outlined, becomes the ease with which the species can be maintained under laboratory conditions. If the species is unusually susceptible to disease or stress under crowded laboratory conditions or otherwise has prohibitive habitat requirements, the difficulties in maintaining a population of test individuals will outweigh the associated problems of monitoring and experimentation.

Extrapolation of Results. The test species selected and the particular response being monitored must have a demonstrated extrapolation to meaningful, real-world environmental processes. Numerous complex systems for the sensing and monitoring of specific biological responses can and have been developed, as evidenced in the biological effects literature. Unless correlations with, or predictions of, recognized biological effects relevant to the survival potential of environmental populations exposed to the particular permutation being tested can be developed, the monitoring method and test species employed have limited utility. These correlations and predictions of survival potential should be developed for numerous chemicals or classes of chemicals potentially reaching surface-water areas in order to lend support to the ultimate utility and overall application of the biomonitoring test species and methods employed.

Cosmopolitan Distribution. The test species selected should have a natural range as wide as reasonably possible in order to lend wide application to results obtained. The selection of local or obscure test species should be avoided since the extension and overall application of data are severely limited. Only in specific cases where monitoring information is required on the effects of a particular environmental variable on an obscure or local species is testing warranted.

Definable Culture Medium and Culture Conditions. It is extremely important to a biomonitoring program in which the test species is used as an integrator of his total environment to be able to adequately define the chemical-physical makeup of that environment. If it is necessary to utilize unreasonably complicated culture conditions to maintain test stocks, an inordinate amount of time is consumed in the maintenance procedures at the expense of biomonitoring and experimentation time. Therefore, it becomes important to the investigator selecting potential test species to consider the amount of laboratory apparatus and time that will be required to maintain continuous stocks of the organism in healthy condition, ready for experimentation at any time.

Trophic Status and Functional Position. Consideration should be given to the trophic status of the potential test species in the real-world environment. Knowledge of the particular habitat and niche is useful in the design of experiments for mode and method of exposure to a test variable. Knowledge of the trophic status allows the investigator to project the results of his biomonitoring experimentation to potential consumers of the test species and develop predictive information for representatives of these additional trophic levels.

Legal Defensibility. The group felt that subsequent legal defensibility of laboratory or field biomonitoring studies would be significantly enhanced through attempts to standardize the test methods and species used. Methods standardization implies acceptability within a defined segment of the scientific community and lends the element of peer review to the program. The process of standardizing a method or approach requires review of the proposed method by certified professionals with qualifications and expertise in that particular area. Thus, when an investigator states he has employed a standard test method or standard approach, he is able to add some degree of quality control to the resultant data by referencing the reviewed and accepted method, enhancing the overall scientific and legal defensibility of his program.

Specific Decision Criteria Indicating the
Need for Biological Monitoring

Monitoring chemical variations in water quality during both laboratory and field investigations is an integral aspect of all pollution monitoring programs. The chemical variables being monitored are typically analyzed from a small water sample taken either at a specific time or as a composite over a specific period. As such, the subsequent analyses performed with that water sample often do not accurately reflect the higher of lower extremes and variations typically seen in levels or concentrations of the chemical parameters in the actual environment being monitored. For example, a dissolved-oxygen reading taken during morning or afternoon hours most probably does not reflect the limiting dissolved-oxygen concentrations with respect to aquatic life since these diurnal minima typically occur during the predawn hours. Similarly, programs designed to monitor concentrations of specific chemical or physical materials (suspended solids, nutrients, trace metals, and organic contaminants) are frustrated by the real-world variables such as seasonal fluctuations in flow, daily and seasonal temperature changes, and significant flow-rate changes following storms. Analyses of specific chemicals and pollutants typically have been subject to these real-world variations, with the result that inherent sample-to-sample variation has been so great as to obscure significant long-term trends.

These inherent variations in chemical concentrations seen in real-world environments strongly imply the need for normalization or integration with respect to time in order to aid interpretation of resultant data. In these instances, an aquatic species continuously exposed to the environment in question serves as the required integrator. Representative aquatic life is strongly tied to the local environment, with only limited capacity to migrate or avoid adverse conditions. Their presence or absence from a particular environment serves to indicate the extremes existing in that environment with respect to chemical-physical variables and pollutants, regardless of the time of day or seasonality of occurrence. Therefore, it is evident that the complex diurnal and seasonal fluctuations influencing ultimate concentrations of an experimental variable in the real world serve in themselves to key the need for a biological monitoring program. Thus, investigators whose field and laboratory programs are attempting to monitor trends in a particular environmental variable potentially have much to gain from the implementation of a biological monitoring aspect to complement and assist interpretation of meaningful environmental impact.

Similarly, investigations designed to assess the impact of complex effluents and toxicity of mixtures have much to gain from the use of aquatic species as response integrators for the particular mixtures. The approach historically taken for the establishment of effluent standards and water quality criteria has been to establish acceptable concentrations based on results of tests and exposures to the singular test material. Rarely is an individual species existing in receiving-water environments exposed to the individual material alone. Much more generally, the test material in question is only one component of a larger array of diverse pollutants seen by the species simultaneously. In these instances, programs designed to assess ultimate impact of test materials should consider the application of biological monitoring techniques to enhance overall application of resultant data to real-world environments.

The Development of Regional, State, and Federal Policies for Biological Monitoring

An obvious consensus was reached by the group that interest in biological monitoring programs was gaining in all facets of academic, industrial, and regulatory agencies. The group felt that the novel nature and research applications of these programs should not be subjected to overly stringent regulatory controls at this time. However, a need was expressed for overall direction with respect to policy formation. Policy descriptions and overall guidelines would be extremely useful to the development and evolution of well-focused biological monitoring programs at all levels of private and public organization. A recommendation was developed by the group for policy direction at the federal government level in order to have greatest potential impact on long-term goals and direction of the numerous biological monitoring applications currently evolving. If such overall policy and direction were offered in the form of recommendations from pertinent branches of the EPA, much original effort currently devoted to programs with little relevance to the real world or to those that appear to be reinventing the wheel with respect to methodology could be eliminated. The group felt that this particular conference and resultant proceedings would certainly assist individuals in an assessment of the state of the art while pointing out the more promising methods and applications of biological monitoring. As in any new research field, the development of methods, subjection to peer review, and ultimate acceptance of these methods by the scientific community would be significantly enhanced and hastened by the early development of commonality in policy and standardization of methods. The development of these methods and their dissemination in this book should encourage their use and application by additional investigators, thus leading to the development of a larger data base including more materials on which the application of biological monitoring techniques can be judged.

22 Issues in Long-Term Biological Monitoring

Ken Dickson and
D.L. Dindal

I. Statement of Major Objectives of Long-term Studies
   A. To instill the resource management approach in answering questions relating biomonitoring to environmental problems; encourage collection of complete data follows:
      1. Ambient habitat information
         a. Location
         b. Presence
         c. Quality
      2. Biotic community structure
         a. Diversity, equitability, richness
         b. Interspecific associations
      3. Biotic community function
         a. Productivity processes
         b. Decomposition
         c. Energy and nutrient fluxes
      4. Temporal and spatial relationships of biotic and abiotic factors within ecosystems
   B. To become fully aware of information redundancy and develop the most efficient methods and monitoring systems
      1. Prepare scientifically valid experimental study designs that are also efficient
      2. Determine implications of redundancy on cost effectiveness
      3. Be aware of both functional and structural redundancy
   C. To gather baseline data and prepare experimental conceptual frameworks as a basis for effective biomonitoring practices
   D. To determine the assimilative capacity of all habitats in question
   E. To distinguish trends attributable to human impacts from natural variabilities and oscillations
   F. To delineate perturbations and compare and relate their impact to various degrees of natural or anthropogenic responses
   G. To relate short-term biomonitoring studies to long-term ones
      1. Long-term monitoring must start with short terms
      2. Establish criteria for long-term intermittent and long-term continuous monitoring

H. To organize, store, and make available data and samples for future reference
   1. Maintain computerized data bank system
   2. Store samples with known characteristics for future assay and comparison
   3. Exchange data and coordinate with the national specimen banking system
I. To develop a statistical perspective for a cost/benefit approach to biomonitoring
   1. Develop the ability to demonstrate differences at the level of precision and power necessary for environmentally meaningful decision making
   2. Evaluate the cost and value of redundancy in monitoring information
   3. Distinguish trends and effects from the noise of natural variability
   4. Relate the results of short-term and acute-effects studies to the prediction of long-term effects and trends

II. Major Areas of Consideration Necessary in Implementing and Solving the Objective Goals
   A. Identification of study area
      1. Obtain specific site for continuous study
         a. Delimit site
         b. Determine how many sites needed to monitor given environmental parameters
      2. Institute permanence of site
         a. Establish stability of ownership
         b. Determine how the permanent monitoring site will be changed by monitoring activities
      3. Collect and record site characteristics
         a. Consider site as a system; describe the total ecosystem
         b. Collect and evaluate background (control) data; establish "background noise"
            (1) Variability within and between background samples
            (2) Temporal and spatial heterogeneity
            (3) Physical-chemical variables; correlation with biological factors
            (4) Flux rates
            (5) Background site processes; cause-and-effect pathways
         c. Delineate calibration sites
            (1) Representative of general control sites
            (2) Maintenance of uniformity
         d. Determine assimilative capacity of control site

B. Personnel-institutional relationships
   1. Establish curricula to teach principles and methods of biomonitoring
      a. Train technicians and professionals for a career in biomonitoring
      b. Provide educators and scientists with necessary background for research and teaching
   2. Nurture interdisciplinary character of biomonitoring
      a. Develop integrated acquaintance with field and laboratory skills;
      b. Stimulate learning of updated knowledge of ecology
      c. Orient environmental thinking into an analysis of systems
   3. Convince universities and government agencies that biomonitoring is or can be a valid research area
      a. Promote national and international value of biomonitoring to all
      b. Remove stigma surrounding biomonitoring versus research
      c. Establish ways in which monitor-related research can count in tenure evaluations
      d. Develop the long-term hypotheses approach
      e. Modify survey, inventory work to include functional data (baseline information on processes) as well as structural data (species lists, standing crop) about biotic communities
   4. Develop cost-effective programs
      a. Educate administrations on the expenses to be expected in long-term programs
      b. Study and develop efficient time-person involvement scheme
C. Necessary development of methods and standards; an attempt to attain a minimal level of competence for all biomonitoring workers
   1. Develop a glossary to standardize all terms and definitions
      a. Prepare for both technical and nontechnical terms
      b. Include ecological terms
         (1) Community structural definitions
         (2) Functional definitions
         (3) Classification of ecosystem components
            (a) Organism
            (b) Population
            (c) Community
   2. Develop and standardize methods for biomonitoring
      a. Structural components
      b. Functional components
         (1) Nutrient cycling
         (2) Productivity, respiration, and utility of P/R ratio
         (3) Secondary productivity
         (4) Microbial-enzymatic assays
      c. Perturbation

        d. Microcosm assays
        e. Synthetic concepts relating structure and function aspects
        f. Appropriate instrumentation
           (1) Field evaluations; chemical, physical, and biological sampling methods
           (2) Microcosm, microbial, and enzymatic assay technology
           (3) Remote sensing
           (4) Computer techniques
           (5) Environmental data base management
        g. Field calibration sites

III. Recommendations for Institutional Policymakers and Program Directors
    A. Policymakers of governmental, educational, and private institutions should recognize the value of long-term biomonitoring to the national interest since biomonitoring provides information on the effects of environmental stress and serves as a basis to evaluate future environmental perturbations.
    B. Educational institutions must recognize the interdisciplinary nature to the solution of environmental problems and thus develop the necessary curricula to meet this end.
    C. Several types of reduced-scale workshops on biological monitoring should be organized and convened in the future; these should be in regional *workshops* and *field of specialization workshops.*
    D. Sufficient funding should be provided to encourage specifically the fulfillment of the above objectives, the implementation of properly trained personnel, and the development and standardization of appropriate methods and procedures required by an effective, national, long-term biomonitoring program.

Index

academic programs, 16, 20, 199, 207, 215-216
accumulators, pollutants, 123-123. *See also* bioaccumulators
acid rain, 117, 139
acute toxicity, effluents, 33, 54, 201, 205
air quality, vegetation as monitors, 117-119, 125, 185
algal growth, studies, 18, 30, 33, 40, 169, 180. *See also* biostimulatory tests
American Institute of Biological Sciences, 7
Ames test for mutagenesis, 169, 205, algal growth response, 17, 33
aquatic bioassays, 25-26, 33, 54, 179; fish, 37, 40-41, 57, 65; macroinvertebrates, 33, 37, 40-41, 50, 94-95. *See also* fish
assimilative capacity of habitats, 213.

bacteria as indicators, 18, 129, 175-178, 199, 201, 215
banking systems, data, 146, 214. *See also* environmental specimen banks
baseline data, 147-148, 213-214
behavioral tests and indices, 69, 75, 191. *See also* fish
benthic communities, 32, 86-87, 91, 97, 100-101, 113-114
Best Available Technology, 12, 20
bioaccumulation, 30, 33, 93, 117, 123
bioaccumulators, 26, 30, 33, 54, 58, 93-94, 124, 181
bioassays of environmental samples, 17, 30 33, 169-180, 199, 215; genotoxic tests, 169; short-term tests, 169-177
biochemical indicators, 32, 125
bioindicators, 32, 147. *See also* bacteria as indicators; benthic communities; biochemical indicators; fish bioindicators of toxic effluent; macroinvertebrates; microbial indicators; mussel watch network; plants (vegetation as monitors); shellfish
biological integrators, 56, 58. *See also* bioaccumulators
biological integrity, 25, 27, 31-33, 97. *See also* community structure and function.
biological monitoring. *See* academic programs; criteria for biomonitoring; definitions; early warning; inventory of biomonitoring programs; legislation; research needs
biological oxygen demand, 91
biosphere reserves, 140-141

biostimulatory tests, 30, 33
biotic index, 97, 104-108. *See also* Shannon-Wiener index
bottle test for algal growth, 33, 180
breathing response in fish, 58, 66, 76. *See also* respiratory rates of exposed fish; ventilation rates

carcinogens, environmental, 33, 94, 173-174
chemical effects tests, 17, 30, 33, 58, 64-65, 75, 83-84, 93-95, 103
chemical mutagens, 33, 169, 172, 174, 185-187
chlorophyll reference materials, 49
chronic bioassays, 33, 66-67, 75. *See also* fish
coastal waters, biomonitoring, 53, 141
Comprehensive River Basin Projects, 35
Community Structure and function, 25-26, 34, 201, 215
continuous vs. intermittent monitoring, 13, 54-57, 201, 204, 213
control data, natural background, 76, 91, 118, 139-140, 151, 198, 214
control vegetation test chambers, air pollutants, 81
costs and benefits, biomonitoring, 16, 145, 214
Council on Environmental Quality, 5-10, 142, 153; Interagency Monitoring Task Force, 5-9
credibility of biomonitoring, 3, 19. *See also* scientific validity
criteria for biomonitoring, 13, 19, 26, 204, 206, 213. *See also* guidelines for biomonitoring

data collection, storage, management, 6, 9, 15, 18, 147, data bank system, 214; data centers, 148; feedback, 9, 56; interpretation, 198; legal use of data, 204
definitions, 5, 12, 29, 137
Department of Energy, ecological monitoring, programs, 142-143; energy-related projects, 163
disseminating information, 149, 153, 195; needs for journal, society, 195
diversity changes, indices, 27-28, 32, 86, 90, 97, 147

early warning, biomonitoring role in, 57, 124, 140

217

ecological effects, 139, 148, 179
ecological monitoring, 7-8, 137-139, 142, 146-149
ecosystem changes, 57, 137, 139, 147-148
educational issues, 20, 113, 149, 161, 197, 207, 215
effluent testing, 15, 17, 30, 58, 83, 103-118, 201, 205
endangered species, 138
EPA., 38, 94, 123, 169
Environmental Monitoring and Support Laboratory, EPA, 35, 39
environmental specimen banks, 147, 214
environmental stress, biomonitoring, 140, 148-149
estuarine waters, 93
Experimental Ecological Reserves, 141

federal biomonitoring, 45, 48-49, 93, 143, 161, 169, 185
Federal Water Pollution Control Act, 26-29, 137
fish, bioindicators of toxic effluent, 18, 57, 69, 75, 84; protection and biomonitoring of, 26-27; and Wildlife Service, 28, 29, 140, 153
funding biomonitoring programs, 140, 142, 144-146
future roles for biomonitoring, 20

genetic effects, 117, 173; point mutations, 173
global biomonitoring, 10, 141, 145
Great Lakes water quality, 31
ground water quality, 31, 199
growth rates, 33, 117, 213. See also biostimulatory tests
guidelines for biomonitoring, EPA., 13, 19, 26

health of environment, DOE programs, 146. See also National Environmental Research Parks
heavy metal uptake, 93, 97, 117, 124

indicator communities, 25, 87, 97, 113, 148, 213, 215
industrial effluent monitoring, 17, 31, 57, 75, 103, 203
information on projects, 151; needs, 195; by states, 162; by subject category, 163
in-plant tests, 33
in-situ tests, 33
institutional relationships, biomonitoring, 20, 43-45, 140-142, 164, 199, 210, 216
integrators of pollutants, 37, 54, 179. See also fish; plants (vegetation) as monitors; shellfish

intermittent vs. continuous monitoring, 213
inventory of biomonitoring programs, 7, 153
in-vitro toxicity, 174. See also Ames test for mutagenesis

legal defensibility, 209. See also scientific validity
legislation, federal, 6, 20, 27-30, 206
long-term biomonitoring, 139, 149, 213

macroinvertebrates, 18, 30, 76, 86-89, 98, 100, 114
management of the environment, 12, 56, 63-64, 11-15
man and the biosphere, 140. See also biosphere Reserves; National Environmental Research Parks
marine organisms, 93-95; ecosystem changes, 29, 93, 141
mathematical models of ecosystems, 148
maximum acceptable toxicant concentrations, 65, 75, 83-84
methodology, 57, 94, 146
microbial indicators, 18, 126, 130, 175, 187, 199, 201, 215
monitoring networks, 34, 36-37, 93, 148
mussel watch network, 93
mutagenesis, 33, 169-175, 185-187

National Biological Monitoring Inventory, 153-154
National Environmental Policy Act, 138, 143-144
National Environmental Research Parks, 146-151
National Heritage Conservation Program, 138
National Institute of Environmental Health Sciences, 186, 191
National Oceanic and Atmospheric Administration, 93, 141
National Pollution Discharge Elimination System, 31, 35, 205
National Science Foundation, 141; proposal for Experimental Ecological Reserves, 141; U.N. proposal for Global Environmental Monitoring System, 141
National Specimen Banking System, 214
natural vs. man-caused environmental stress, 137-140, 146, 198, 214
neurological-behavioral indicators, 58, 191
nutrient pollution, 31-33, 213, 216. See also algal growth, biostimulatory tests

Oak Ridge National Laboratory, 153; biomonitoring inventory, 153; information dissemination, 195-196

Index

occupational environments, 194, 201

pathways and sinks for pollutants, 56, 149, 214
physiological response to toxic chemicals, 64-65, 79, 205
plankton indicators, 18, 30-33
plants (vegetation) as monitors, 117, 125, 186
point discharges, 57
policy (state-federal), 5, 20, 27-29, 137, 142, 145, 210, 216
predictive models, 64, 83, 93, 148, 175, 198, 205
productivity, propagation, 27

quality assurance, 9

rates of biomonitoring, 3, 7, 13-15, 54, 117, 123, 146, 205; ambient monitoring, 25
real-time monitoring, 54, 209
receiving waters, 11, 19, 28-30, 57, 206; assimilative capacity, 56, 213
redundancy in monitoring data (costs and benefits), 214
regional EPA programs, 37, 210
regulatory programs using biomonitoring, 15, 25-29, 35, 201, 206
remote sensing, 83, 138, 216
research needs, 9, 113-114, 118, 197-199, 201
resilience of ecosystems, 140, 148, 213
respiratory rates of exposed fish, 64. *See also* breathing response in fish

saprobien indicator system, 98
scientific validity, 16, 19, 140, 174, 213
sediment populations, 95, 101, 201. *See also* benthic communities, 101
Shannon-Wiener index, 86, 97, 104, 107-108

shellfish, 27-28, 93, 124
short-term bioassays, 59, 175; "core" battery genotoxic tests, 171
soil biomonitoring, 124-130, 181
special composition, 31
specimen banks, 153
standard methods, 16-18, 45, 113, 201, 207, 215
states biological monitoring, 43-44, 46, 161-163
statistical techniques, 18, 79-80, 113, 214
status monitoring, 138
stream surveys, 85
synergistic effects, 58, 104, 148, 154

taxonomic composition, 18, 26, 34, 114; projects by taxa categories, 162
temperature impact, 30-31, 101
teratogens, 33, 172, 177
thermal water discharges, 31. *See also* temperature impact
tolerance of bioindicator to pollutants, 19, 90, 113
tolerance to pollutants, 19, 48, 90, 99, 113
toxicity testing, overview, 102, 205; chronic effects, 166
Toxic Substances Control Act, 64, 65
tradescantia-mutagen indicator, 186
trends, biomonitoring studies, 138

vegetation and air/soil monitors, 117-121, 124-129; growth and yield, 117
ventilation rates, as index of fish exposure, 79-82. *See also* breathing response in fish; respiratory rates of exposed fish

waste discharge monitoring, 10, 15, 26-31, 57, 94, 100
watershed management, 10, 15, 19, 28, 54, 84

About the Contributors

Donald V. Bradley received the M.S. degree in biology in 1968 from the University of Nevada. He was employed as staff research assistant at the EPA Environmental Monitoring and Support Laboratory from 1968 to 1978. His principal area was the development of biological monitoring techniques for assessing the effects of environmental pollution on vegetation and soil organisms.

John Douglas Buffington received the Ph.D. in biology at the University of Illinois in 1965. Currently, he is a senior scientist with the Council on Environmental Quality, Executive Offices of the President. In this position, Dr. Buffington's responsibilities include directing the activities of interagency committees that recommend appropriate national environmental monitoring programs to the president and Congress.

Robert L. Burgess received the Ph.D. in ecology from the University of Wisconsin. He is head of the environmental resources section in the environmental sciences division at Oak Ridge National Laboratory. Since 1975 he has been an adjunct professor of ecology at the University of Tennessee. Dr. Burgess is a past president of the North Dakota Academy of Science and the North Dakota Natural Science Society and has held numerous positions in the Ecological Society of America.

John Cairns, Jr., is a professor in the Biology Department and director of the Center for Environmental Studies at Virginia Polytechnic Institute and State University. He has been a leader in developing standardized biological monitoring methods for measuring the adverse impact of environmental pollutants on aquatic organisms. Dr. Cairns has contributed to numerous publications on his research in using biological indicators to measure and evaluate effects of human activities on biological systems.

Arthur W. Cooper received the B.A. and M.A. degrees in botany from Colgate University and the Ph.D. degree in botany from the University of Michigan.

He is professor and head of the Department of Forestry, North Carolina State University. Dr. Cooper is president of the Ecological Society of America. His past research interests have dealt with plant ecology, plant community studies, estuarine ecology, primary productivity in salt marshes and forests, and application of ecology to land-management problems.

John A. Couch received the Ph.D. from Florida State University and received postgraduate training in public health at The Johns Hopkins University and the Armed Forces Institute of Pathology. He is an adjunct professor in biology at the University of West Florida and is presently task leader in pathobiology and team coordinator for carcinogen research for the EPA Environmental Research Laboratory in Gulf Breeze, Florida.

Ford A. Cross received his graduate training in oceanography at Oregon State University in 1968. He has worked at the National Marine Fisheries Service, Beaufort Laboratory, Beaufort, North Carolina, since 1967 and is currently chief of the division of estuarine and coastal ecology. Dr. Cross manages a research program designed to determine the relationship between estuarine and fishery productivity and the impact of human activities on fishery resources.

Kenneth Dickson received the Ph.D. in zoology from Virginia Polytechnic Institute and State University in 1971. He was an associate professor of zoology and held a concurrent position as assistant director for the Center of Environmental Studies at the university from 1970-1978. Presently, in addition to his teaching and research activities, Dr. Dickson is coordinating environmental academic and research programs as the director of the Center for Environmental Studies at North Texas State University.

Daniel Lee Dindal received the B.S. in wildlife management and the M.S. and Ph.D. in ecology from Ohio State University. He is a professor of ecology at the State University of New York School of Forestry. His principal interests include the ecology of soil invertebrates in natural and manipulated terrestrial microcommunities.

David Gruber received the M.S. degree in marine sciences from Long Island University and received the Ph.D. in 1976 from Texas A&M University for his work in chemoreceptive capabilities of sharks. Dr. Gruber is a research associate at the Center for Environmental Studies at Virginia Polytechnic Institute and State University in Blacksburg. His work involves developing a biological monitoring concept for automated and continuous on-line assessments of waste-water discharges.

Walter W. Heck received the Ph.D. in botany from the University of Illinois in 1954. He has been a research leader and technical advisor in air-pollution research in the mid-Atlantic area, southern region, Science Education Administration/Agriculture Research (SEA/AR), and a professor of botany at North Carolina State University, Raleigh, since 1973. His current research includes studies on the use of plants as sensitive indicators and monitors of air pollutants. He is also participating in a major effort at North Carolina State University to evaluate the effects of acid rain.

Alan Hirsch received the B.S. and M.S. degrees in zoology from Michigan State University and the Ph.D. in conservation from the University of Michigan. Prior to his present assignment as deputy assistant administrator for environmental processes and effects research, Environmental Protection Agency, he served from 1974 to 1979 as a senior ecologist and chief of the Office of Biological Services, U.S. Department of the Interior.

Herbert C. Jones III received the B.S. in forest management in 1960 and the Ph.D. in botany in 1965 from the University of Florida. He has twelve years of experience in monitoring of and research on the effects of atmospheric emission from coal-fired power plants and fertilizer plants on air quality and crops and forests. Presently Dr. Jones is supervisor of the air quality research section, air quality branch, Office of Natural Resources, of the Tennessee Valley Authority in Muscle Shoals, Alabama. Dr. Jones is responsible for supervising research on the dispersion, long-range transport, chemistry, and fate of coal-fired power-plant pollutants.

D.R. Lena received his graduate training in biology from the University of North Carolina at Chapel Hill. He now works with the North Carolina Division of Environmental Management. He is presently concerned with the use of fresh-water benthic and zoo plankton as indicator organisms to determine the effects of environmental stress.

Linda West Little has been president of L.W. Little Associates, Environmental Consultants, since 1979. She has served as an environmental consultant with the Atomic Safety and Licensing Board Panel of the U.S. Nuclear Regulatory Commission since 1974. Dr. Little was an associate professor of environmental biology, Department of Environmental Sciences and Engineering, University of North Carolina at Chapel Hill, from 1974 to 1977 and is now an adjunct professor with that department.

Frank G. Lowman received the Ph.D. in fisheries from the University of Washington, Seattle, in 1956. Dr. Lowman has served as associate director in environmental sciences and chief scientist and head of the Marine Biology Division at the Puerto Rico Nuclear Center. He is currently a senior biologist with the EPA Environmental Research Laboratory at Narragansett, R.I., where he is actively engaged in managing the laboratory's mussel-watch network.

J.C. McFarland received the B.S. degree in plant physiology in 1971 from the University of California at Riverside. He has been employed by the Environmental Protection Agency since 1973.

Alan W. Maki is the director of the environmental safety department for the Ivorydale Technical Center, The Proctor and Gamble Company, Cincinnati, Ohio. Dr. Maki has been a leader in the scientific community in developing reliable, standardized, and sensitive biological monitoring techniques for measuring the effects on aquatic organisms of industrial and community effluents.

Clifford L. Mitchell received the Ph.D. degree from the University of Iowa in 1959. He is chief of the Laboratory of Behavioral Toxicology at the National Institute of Environmental Health Sciences, Research Triangle Park, North Carolina.

John C. Morse received the Ph.D. in entomology in 1974 from the University of Georgia. He is currently an associate professor of entomology and taxonomy at Clemson University, South Carolina.

William S. Osburn, Jr. received the Ph.D. in botany from the University of Colorado in 1958. He served as associate professor of radiation biology at Colorado State University in 1965-1966. In 1966 he was appointed to a position as ecologist with the division of biology and medicine of the Atomic Energy Commission and has continued to serve in this capacity with the Department of Energy Office of Environmental Programs.

David Penrose received the M.P.H. degree from the University of Michigan in 1975. He is employed as an environmental biologist with the North Carolina Division of Environmental Management. His research interests include the taxonomy and ecology of aquatic macrobenthos.

R.C. Rogers received the B.S. in soil microbiology from North Carolina State University in 1974. He has been a staff biologist with the Environmental Protection Agency in Las Vegas since August 1974. His principal interest is developing techniques to assess the impact of environmental pollution on selected species and in populations of microorganisms in the soils.

Michael D. Shelby received the B.S. in biology from Central State College, Edmond, Oklahoma, in 1966 and the Ph.D. from the University of Tennessee, Knoxville in 1973. He has worked at Oak Ridge National Laboratory in the establishment and operation of two computerized toxicology information centers, the Environmental Mutagen Information Center (EMIC) and the Environmental Teratology Information Center (ETIC). Currently serving as the assistant to the associate director for genetics at the National Institute of Environmental Health Sciences, Dr. Shelby is involved in a variety of projects dealing with environmental mutagenesis and short-term tests for carcinogenicity.

L.A. Smock received the Ph.D. in biology from the University of North Carolina at Chapel Hill. He is now a faculty member in the Department of Biology at Virginia Commonwealth University. His research interests include studies on the structure and trace-metal cycling in benthic communities.

James Stewart received the B.S. in agronomy and the M.Ed.M. and Ph.D. in adult education from North Carolina State University. Currently he is acting director of the North Carolina Water Resources Research Institute. Dr. Stewart has been active nationally in technology transfer with the Office of Water Research and Technology. In 1979, he was recipient of the North Carolina State University Outstanding Extension Service Award.

Joab L. Thomas received the B.S., M.A., and Ph.D. degrees in biology from Harvard University. He was a member of the research and teaching staff of the Arnold Arboretum of Harvard University from 1959 to 1961. Dr. Thomas has been chancellor of North Carolina State University at Raleigh since 1976. He has published books, monographs, and papers in the fields of botany and higher education.

Michael D. Waters received the Ph.D. degree in biochemistry from the University of North Carolina in 1969. He now serves as coordinator of the genetic toxicology program and chief of the biochemistry branch, Environmental Toxicology Division of the Environmental Protection Agency at Research Triangle Park, North Carolina. Dr. Waters's research involves the use of microbial, cell, and organ-culture techniques as well as intact animals in the study of genetic, biochemical, and physiological effects of environmental pollutants at the cellular and subcellular levels.

Cornelius Weber received the Ph.D. in plant pathology from Iowa State University in 1966. For the past thirteen years he has had supervisory federal assignments in research to develop methods for measuring the biological effects of environmental pollution on communities of aquatic organisms. Dr. Weber is presently chief in the Aquatic biology section of the EPA Environmental Monitoring and Support Laboratory in Cincinnati.

Bruce Wiersma received the Ph.D. from the State University of New York School of Forestry at Syracuse in 1968. Dr. Wiersma has been with the EPA since 1970. Since 1974 he has served as a staff ecologist with the EPA Environmental Monitoring and Support Laboratory located in Las Vegas. His principal interests are in developing sensitive techniques for measuring trace environmental contaminants and assessing their biological effects.

About the Editor

Douglas L. Worf received the B.S. degree in chemical engineering from the University of Toledo in 1938 and the Ph.D. in biochemistry from Georgetown University in 1953.

Dr. Worf has been active in planning and developing environmental research programs and policy for several federal agencies and for the Executive Offices of the President. During the period 1965–1967 he was the research director for the Alaska Water Laboratory, a federal facility located in College, Alaska. Prior to this he was a staff scientist with the Division of Biology and Medicine, Atomic Energy Commission, where he was responsible for developing and managing research and monitoring directed at assessing radioactive levels and biological effects from various atomic energy activities.

Dr. Worf was appointed as part-time visiting professor in environmental studies at North Carolina State University and is now a consultant and policy advisor for several organizations concerned with environmental quality, energy alternatives, chemical safety, and toxicology.